# XIX International Youth Scientific School on Coherent Optics and Optical Spectroscopy 2015



Journal of Physics: Conference Series !!!!!!!!!!!!!!!!
Volume 714

Kazan, Russia
5 – 7 October 2015

**Editors:**

**Salakhov Myakzyum Khalimullovich**
**Samartsev Vitalii Vladimirovich**
**Gainutdinov Renat Khamitovich**

ISBN: 978-1-5108-2453-9
ISSN: 1742-6588

**Printed from e-media with permission by:**

Curran Associates, Inc.
57 Morehouse Lane
Red Hook, NY 12571

proceedings.com

**Some format issues inherent in the e-media version may also appear in this print version.**



Printed by Curran Associates, Inc. (2016)

For permission requests, please contact the Institute of Physics
at the address below.

Institute of Physics
Dirac House, Temple Back
Bristol BS1 6BE UK

Phone: 44 1 17 929 7481
Fax: 44 1 17 920 0979

techtracking@iop.org

**Additional copies of this publication are available from:**

Curran Associates, Inc.
57 Morehouse Lane
Red Hook, NY 12571 USA
Phone: 845-758-0400
Fax: 845-758-2633
Email: curran@proceedings.com
Web: www.proceedings.com

# Table of contents

## Volume 714

**XIX International Youth School on Coherent Optics and Optical Spectroscopy5–7 October 2015, Kazan, Russia**

**Accepted papers received: 22 April 2016 Published online:**

### Preface



### Papers

The XIX International Youth School on Coherent Optics and Optical Spectroscopy (COOS2015) was held in Kazan, Russia, from October 5 to October 7 at the Nikolai Lobachevsky Scientific Library of Kazan Federal University. The School follows the global tendency toward comprehensive studies of matter properties and its interaction with electromagnetic fields. Since 1997 more than 100 famous scientists from USA, Germany, Ukraine, Belarussia and Russia had plenary lecture presentations. This is the right place, where over 1000 young scientists had an opportunity to participate in hot discussions regarding the latest scientific news. Many young people have submitted interesting reports on photonics, quantum electronics, laser physics, quantum optics, traditional optical and laser spectroscopy, non-linear optics, material science and nanotechnology. Here we are publishing the full-size papers prepared from the most interesting lectures and reports selected by the Program Committee of the School. Plenary sessions were offered by the following invited speakers:

**Ildar Gabitov**, University of Arizona, USA.

- High speed fiber optic telecommunications

**Andrei Naumov**, Institute for Spectroscopy RAS, Troitsk, Moscow, Russia.

- The manifestation of local field effects in the spectra of single impurity molecules in solids

**Askhat Basharov**, National Research Center 'Kurchatov Institute', Moscow, Russia.

- A simple way of introducing stochastic differential equations
- Generation of terahertz radiation in the interaction of coherent electromagnetic wave with a resonant medium in a constant external electric field

**Anastas Bukharaev**, Kazan E. K. Zavoisky Physical -Technical Institute, Kazan, Russia.

- Investigation of strain-induced magnetization change in ferromagnetic. Review

**Andrei Maimistov**, National Research Nuclear University MEPhI, Moscow, Russia.

- Optical flat bands in 2D waveguide arrays with alternating sign of refraction index

**Sergey Sazonov**, National Research Center "Kurchatov Institute", Moscow, Russia.

- A simple way of introducing stochastic differential equations

**Vitalii Samartsev**, Kazan E. K. Zavoisky Physical -Technical Institute, Kazan, Russia.

- Femtosecond correlated transient holography in CdS crystal under two-quantum excitation

**Vasily Klimov**, P.N. Lebedev Phsical Institute of RAS, Moscow, Russia.

- New optical properties of nano-holes and their applications

**Vitalii Samartsev**, Kazan E. K. Zavoisky Physical -Technical Institute, Kazan, Russia.

- Femtosecond correlated transient holography in CdS crystal under two-quantum excitation

**Elena Kazantseva**, National Research Nuclear University MEPhI, Moscow, Russia.

- Solitary wave generation from constant continuous wave in nonlinear waveguide coupler which consists of two waveguides with positive and negative refraction indices

**Andrei Gaidash**, ITMO Unifersity, Saint Petersburg, Russia.

- Counter-attack by means of photon beam splitting in quantum cryptography with a multi-photon state detector

**PROGRAM COMMITTEE**

Gabitov Ildar Ravilievich – co-chairman
Dr. Sc. (Phys.-Math.), Professor at University of Arizona, USA
Naumov Andrey Vitalievich - co-chairman
Dr. Sc. (Phys.-Math.), Professor, deputy director at Institute for Spectroscopy RAS
Salakhov Myakzyum Khalimullovich
Dr. Sc. (Phys.-Math.), academician of the Tatarstan Academy of Sciences, president the Tatarstan Academy of Sciences
Samartsev Vitalii Vladimirovich
Dr. Sc. (Phys.-Math.), academician of the Russian Academy of Natural Science, Professor at Kazan E. K. Zavoisky Physical -Technical Institute ( KPhTI )
Gainutdinov Renat Khamitovich
Dr. Sc. (Phys.-Math.), professor at Kazan Federal University

**ORGANIZING COMMITTEE**

Salakhov Myakzyum Khalimullovich – rector of the School Samartsev Vitalii Vladimirovich – vice-rector of the School Kamalova Dina Ilevna – scientific secretary of the School

This proceedings volume contains papers reported at the conference. The Proceedings of the XIX International Youth School on Coherent Optics and Optical Spectroscopy (COOS2015) were edited by Salakhov Myakzyum Khalimullovich, Samartsev Vitalii Vladimirovich and Gainutdinov Renat Khamitovich with the assistance of Petrova Alexandra Sergeevna as copy editor.

We are grateful to The Optical Society of America, Russian Foundation for Basic Research, The non-profit Dynasty Foundation, Tatarstan Academy of Science, The Ministry of Education and Science of The Russian Federation for financial.

**Salakhov Myakzyum Khalimullovich, Editor**
**Samartsev Vitalii Vladimirovich, Editor**
**Gainutdinov Renat Khamitovich, Editor**

# Peer review statement

All papers published in this volume of *Journal of Physics: Conference Series* have been peer reviewed through processes administered by the proceedings Editors. Reviews were conducted by expert referees to the professional and scientific standards expected of a proceedings journal published by IOP Publishing.

# FTIR spectroscopy of conformational probes introduced into fullerene-containing branched poly (methyl methacrylate)

**L R Abdrazakova and D I Kamalova**

Kazan Federal University, Kazan 420008, Russian Federation

E-mail: dina.kamalova@kpfu.ru

**Abstract**. In this paper we consider the effect of fullerene on the local molecular dynamics of the branched poly (methylmethacrylates). The temperatures of the secondary relaxation transitions were determined by FTIR spectra. It was shown that there is the secondary relaxation transition at 190 K in copolymer containing $C_{60}$, while there are the relaxation transitions at 230 and 190 K in same copolymer without fullerene.

## 1. Introduction

Branched polymers have unique properties: high solubility in non-aqueous media, low viscosity as compared with linear polymers, high sorption capacity and ability to transport low-molecular substances. A linear (nonbranched) poly (methyl methacrylate) (PMMA) is widely used in mechanical engineering, instrumentation, medical industry, household products, and branched copolymers of methyl methacrylate are used as nanocontainers of functional additives in the synthesis of the polymers with new properties. The study of the structure and internal dynamics of branched polymers is relevant and contributes to the development of modern ideas about the fundamental physical properties of macromolecules.

Branched polymers have a large number of lateral groups and compact globular structure in solutions as compared with the linear ones. Due to the unusual topology, branched polymers have good compatibility with different polymers, excess free volume and ability to absorb substances inside macromolecules according to the "host-guest" type.

The addition of fullerene $C_{60}$ as a nanomodifier to branched polymers by covalent and noncovalent bonding leads to a new complex of useful properties and, thus, opens further prospects of its application as new materials with electrical, magnetic and optical properties [1]. Along with diamond, graphite and carbon nanotubes fullerene is allotropic form of carbon. The main advantage of fullerene are its high electron-acceptor properties and the ability to synthesize various chemical derivatives. It is mostly used in pharmacology. A series of derivatives of fullerene $C_{60}$ with pronounced membranotropic, antioxidant, antiviral and immunomodulatory properties has been developed. This has opened the way to the use of fullerene for effective therapies and vaccines [2, 3]. It is known that fullerene is an effective nanomodifier; its small additive significantly improves the operational properties of different polymers [4].

The purpose of this work is IR spectroscopic study of the local molecular dynamics in glassy branched poly (methylmethacrylate) (BPMMA) and its fullerene-containing analogue (FPMMA) by the method of conformational probes [5]. This method is based on investigating conformational behavior of a low-molecular-weight compound introduced as a probe into an amorphous glassy polymer. The value of the equilibrium constant $K$ is determined by the expression:

$$K = \exp(-\Delta H_0 / RT + \Delta S_0 / R),$$

where $\Delta H_0$ and $\Delta S_0$ are the enthalpy and entropy differences of the conformers. The values $\Delta H_0$ and $\Delta S_0$ depend on intramolecular interactions in the probe molecule and dynamic properties of the medium (e.g. polymer). Hence, the information about structure and local mobility of polymers can be obtained by investigating thermodynamic and kinetic parameters of the probe molecule conformations.

A decrease of the temperature causes a redistribution of conformer's concentrations, i.e. a shift of conformational equilibria. Consequently, the intensities of conformer IR-bands are redistributed. According to the Beer's law, one may write the following expression for the integrated intensities $I_t$ and $I_g$ of the absorption bands of the *trans* and *gauche* conformers:

$$\ln(I_t / I_g) = \ln(\alpha_t / \alpha_g) - \Delta H_0 / RT + \Delta S_0 / R = \ln K + \ln(\alpha_t / \alpha_g),$$

where $\alpha_t$ and $\alpha_g$ are the integral absorption coefficients of the utilized bands. If there are conformational transitions in the system and $\Delta H_0 \neq 0$ then the experimental value $\ln(I_t/I_g)$ must depend on the temperature. The $\Delta H_0$ value characterizes a slope of this dependence. If a low-molecular-weight compound is added to a polymer as a probe, then the values $\Delta H_0$, $\Delta S_0$ will also depend on macromolecular structure and local molecular mobility of polymer. If at a certain temperature, $T_f$, the conformational transitions freeze, it is definitely due to a process taking part in the polymer. Comparing the structure of the rotational fragments of probe and the polymer, one may get an idea about the type of secondary relaxation transitions [5].

## 2. Experimental

Branched copolymer of methyl methacrylate (MMA) with allyl methacrylate (AMA), which was obtained under conditions of the transmission of chain (chain transfer agent - 1 decantiol, DT) was investigated. The composition of the initial monomeric mixture of MMA / AMA / DT is equal to 100: 5: 5 mol %. Methyl methacrylate copolymer of the same composition but containing $C_{60}$ was investigated too. The glass transition temperatures are 347 and 340 K for BPMMA and FPMMA, respectively.

We used the conformationally-inhomogeneous probes: 1,2-diphenylethane (DPE); 1,2-dichloroethane (DCE); 1,1,2,2-tetrabromoethane (TBE) which exist in two conformations in liquids and solutions. The volumes $V_p^{\neq}$ of small rotational fragments of the probe molecules estimated by an additive scheme are the following: 24 (DCE), 57 (TBE) and 86 (DPE) Å$^3$. Analytical absorption bands were the following conformation-sensitive absorption bands of probes at 616, 638 and 585 cm$^{-1}$ for TBE; at 655 and 710 cm$^{-1}$ for DCE, at 580 and 700 cm$^{-1}$ for DPE. These bands are located in the transparent region of the polymers.

Films were prepared by dissolving a polymer in chloroform. The solutions were spread on KBr plates. The films were dried at room temperature for several days to evaporate the solvent completely. The absence of solvent in the samples was fixed by IR spectra. We used the cryostat Specac cooled by liquid nitrogen. Fourier-transform infrared spectra were recorded with a Perkin Elmer FTIR spectrometer Frontier.

## 3. Results and discussion

The FTIR spectra of the probe+polymer system for three different temperatures are represented in figure 1. These spectra demonstrate a change of intensities of the conformation-sensitive (gauche and trans) absorption bands of the probe in the polymer when temperature decreases. Similar spectra were obtained for the other systems: probe+BPMMA and probe+FPMMA. The dependencies of logarithm of the ratio of the intensities of the conformationally-sensitive absorption bands upon the inverse temperature were plotted (figure 2). Such dependencies for all probes in branched copolymers consist of two branches which cross at the freezing temperature of conformational equilibrium of the probe in the polymer ($T_f$). The values of obtained temperatures for BPMMA are 195±10 K (DCE), 230±10 K (TBE) and 225±20 K (DPE). The values of temperatures $T_f$ for FPMMA are 190±15 K (DCE), 190±10 K (TBE) and 205±10 K (DPE). The freezing temperature of conformational equilibrium of the probe in the polymer reflects the secondary relaxation transition in this polymer.

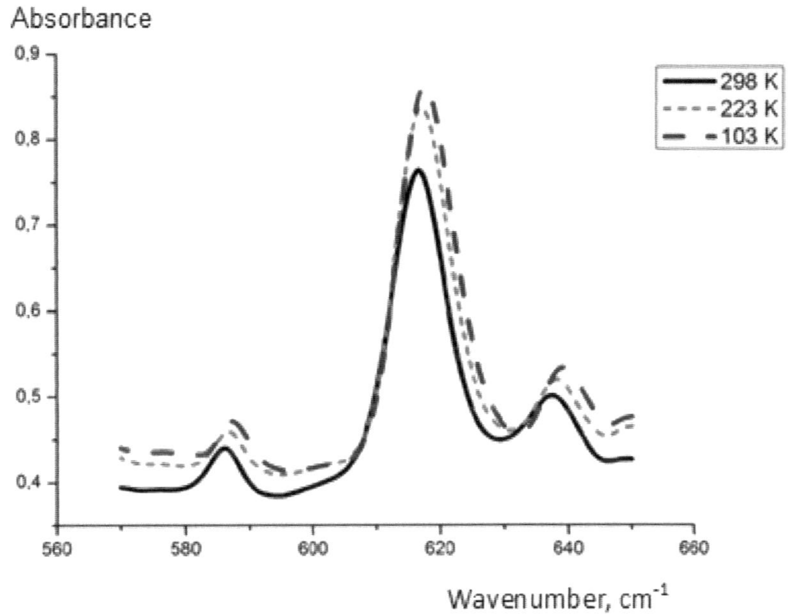


**Figure 1.** Fragments of FTIR spectra of TBE in BPMMA at different temperatures.

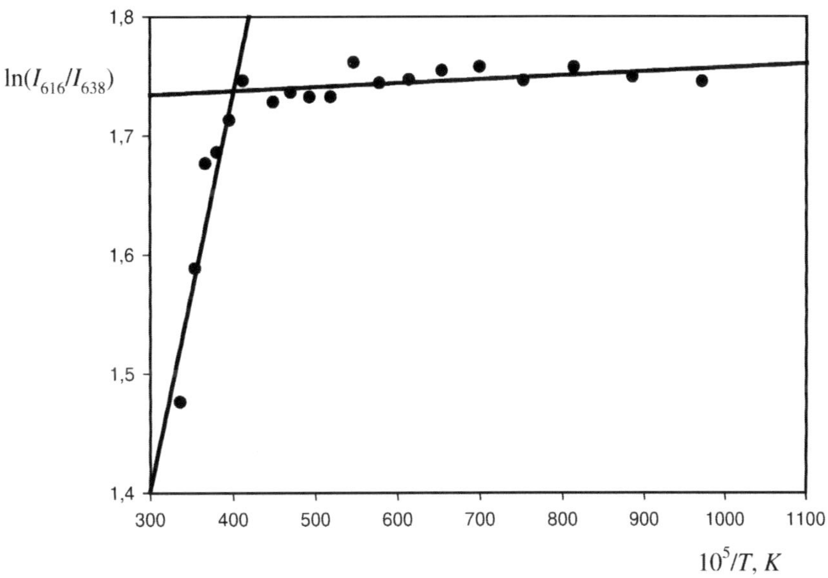


**Figure 2.** Dependence $\ln(I_g/I_t)$ on $10^5/T$ for TBE in BPMMA.

Freezing of conformational mobility of the probe is related to the average size of the cavity in the polymer, which is close to the size of the rotational fragment of the probe molecule ($V_p^{\neq}$). Comparing of the freezing temperatures of conformational mobility of the probes in branched and linear polymers indicates that in branched copolymers BPMMA and FPMMA there is the same relaxation transition ($T_f$ = 190 K) as linear PMMA. Besides, the relaxation transition ($T_f$ = 230 ± 10 K) occurs in BPMMA approximately as in linear PMMA (240 ± 10 K). When we used all three conformational probes having different activation volumes (from 24 to 86 Å$^3$) we have obtained only one relaxation transition in the copolymer with fullerene. Fullerene molecules consist of 60 carbon atoms and have the compact spheroidal shape. Each fullerene molecule occupies very large volume (approximately 190 Å$^3$) as compared with the volume of activation of conformational probes. Covalently bonded fullerene exerts loosening influence on branched copolymer. This influence is such that the conformational mobility of the largest fragment of the used probe is frozen at the lowest temperature.

The authors thank S.V. Kurmaz (Institute of Problems of Chemical Physics of the Russian Academy of Sciences) for the provision of polymer samples.

**References**

[1] Kurmaz L V and Ozhiganov V V 2011 *High-molecular connections* **53** 394-409
[2] Anufrieva E V and Krakowiak M G 2002 *Solid State Physics* **44** 443-444
[3] Salikhov T H , Tabarov S H , Rashid D A *et al* 2009 *Letters to the Journal of Technical Physics* **35** 75-78
[4] Zgonnik V N and Vinogradov L V 2002 *Solid State Physics* **44** 592-593
[5] Kamalova D I *et all* 2008 *Konformatsionnie zondi v izuchenii lokalnoi podvizhnosti polimerov* (Moscow: Fizmatkniga)

Journal of Physics: Conference Series **714** (2016) 012002 doi:10.1088/1742-6596/714/1/012002

# NOT and CNOT gates for photon logical qubits on different frequency states

**M Akbari[1], S N Andrianov[2] and A A Kalachev[1,2,3]**

[1] Kazan Federal University, 18 Kremlyovskaya St, Kazan 420008, Russia
[2] Institute for Perspective Research of Tatarstan Academy of Sciencies, 36a Levobulachnaya St, 420011, Kazan, Russia
[3] Zavoisky Physical-Technical Institute, 10/7, Sibirsky tract, 420029, Kazan, Russia

E-mail: a.a.kalachev@mail.ru

**Abstract.** Starting from Hamiltonian of three level atom interacting with photons, we calculate effective Hamiltonian by Schrieffer-Wolff transformation. We obtain the operation of photon frequency conversion solving appropriate Schrodinger equation. Based on this operation, an efficient protocol is proposed for implementing quantum NOT and CNOT gates for applications in quantum information processing.

## 1. Introduction
Now we know that non-classical character of light, such as entanglement and anti-bunching, is conserved during the process of quantum frequency conversion [1-4]. If some of the nonlinear optical processes are accomplished using only one photon it would be advantageous; these single-photon devices are anticipated to be more robust than those requiring few photons for the same nonlinear processes. Thus far, for single-photon frequency conversion an abundant variety of methods [1, 2, 5–19] is provided, videlicet multiwave- mixing [1, 2, 5–7, 13–17, 19], cavity opto-mechanics system [9, 10] and single-photon adiabatic wavelength conversion [11]. Recently, a novel realizable real-space theoretical approach and pseudospectral numerical method with near unity conversion efficiency are proposed for single-photon frequency conversion via constructive interference in Sagnac interferometer coupled to a three-level atom [12, 13]. They show that high frequency conversion at a single photon level can be achieved due to the interference when the coupling strengths between different atomic transitions to the waveguide loop of the Sagnac interferometer are equal, and additionally with resonant condition which has fixed the photon frequencies. Single-photon frequency up-conversion is employed to change a telecom-band photon into a visible wavelength photon so that single-photon detector technologies for the visible wavelength bands can be utilized [20]. Quantum frequency conversion is widely used in high-sensitivity detection of weak optical signals [21-23]. The capability of converting photons from one wavelength to another wavelength is a key requirement for combining the photons in telecommunications band for quantum transmission and the photons in near-visible band for quantum storage. It has many potent applications in quantum

communication and quantum information processing [23-30]. Our results pave the way for implementing quantum NOT and CNOT gates using frequency conversion of photons on a three-level atom.

The paper is organized as follows. In section 2 we review single photon frequency conversion and in section 3, we represent effective Hamiltonian via Schrieffer-Wolff transformation and our theoretical approach based on this approach results in an exact formula for the general state of the system. In section 4, we present an analytical result, including the method of single-photon frequency conversion for constructing NOT and CNOT gates.

## 2. A physical standpoint of the frequency-conversion process

The basis of the system of interest shown in figure 1 is a $\Lambda$ -type three-level gate atom with energy level structure shown in figure 2. The atom directly couples to the Sagnac waveguide loop, and an input photon can move from the waveguide into the Sagnac loop through a 50:50 beam splitter. When the atom initially in the $|x\rangle$ state is excited by an incoming photon, it can spontaneously decay to either of the low-lying states $|x\rangle$ or $|y\rangle$ and emit a photon.

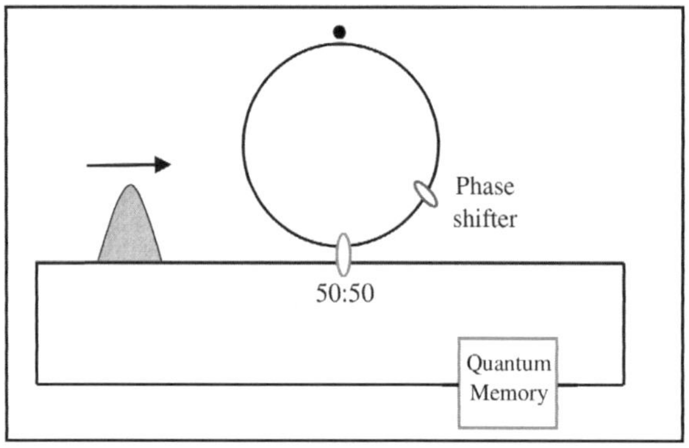


**Figure 1.** Schematic diagram of a Sagnac interferometer coupled to a $\Lambda$-type three-level atom and waveguide modes. Sagnac loop is connected with a waveguide with a 50:50 BS. In the Sagnac loop there is a tunable phase shifter which can be used to control the relative phase between the cw rotating photons and ccw rotating photons.

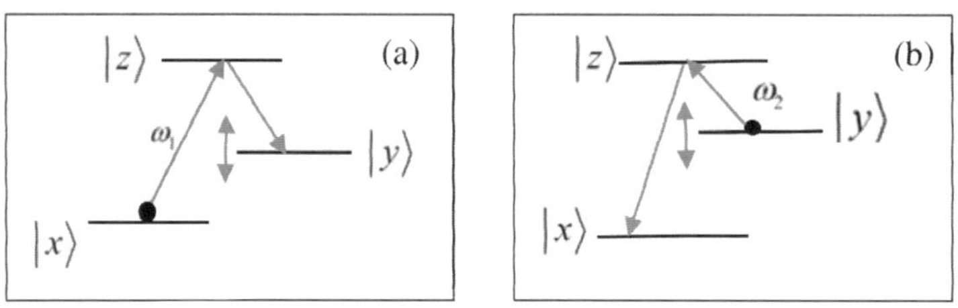


**Figure 2.** Energy level structure of the gate atom in the case of frequency down-conversion (a) and up-conversion (b). The $|x\rangle$ and $|y\rangle$ are the ground states, and $|z\rangle$ is the excited state. The atomic level $|y\rangle$ is shifted via Stark effect.

If the atom decays to state $|y\rangle$, the outgoing photon will be frequency downshifted [see figure 1(a)]. Note that frequency up conversion can also be achieved if the atom is initialized to the state $|y\rangle$ and decays to the state $|x\rangle$ [figure 1 (b)]. If an incoming photon encounters the atom directly, the outgoing photon will be in a spectral superposition of frequency-shifted and -unshifted waves (i.e. a spatial superposition of reflected and transmitted waves), with a maximum frequency-shift probability of one half. The conversion efficiency can be significantly enhanced by exploiting quantum interference: if the input photon is in a spatial superposition of incoming from the left and incoming from the right, emission from the $|x\rangle \leftrightarrow |z\rangle$ transition (in the down-conversion case) or the $|y\rangle \leftrightarrow |z\rangle$ (in the up-conversion case) can be completely suppressed through quantum interference. That is, the atom undergoes a complete population state transfer, induced by the input photon. The input photons propagate in opposite directions of the Sagnac interferometer and constructively interfere at the position of the atom. Constructive interference suppresses the worrisome transition to improve the efficiency of frequency conversion. In another hand, the superposition of the clockwise and counterclockwise moving states can be prepared in the waveguide loop. In certain cases, the interference resulting from the superposition has a constructive effect on the inelastic scattering and a destructive effect on the elastic scattering. The frequency-shift probability reaches its maximum when the coupling strengths $g_{xz}$ and $g_{yz}$ are equal. In addition, in the Sagnac waveguide loop, a phase shifter (as shown in figure 2) can be used to adjust the relative phase $\theta$ between the clockwise and counterclockwise propagating photons. When the relative phase is zero (i.e., $\theta$ = 0), the input photons from two paths will constructively interfere and form the spatial superposition at the position of the atom.

## 3. The effective Hamiltonian and exact solutions of the Schrodinger equation

The Hamiltonian of the system is $H = H_0 + H_1$; where $H_0$ and $H_1$ are unperturbed and perturbed parts respectively:

$$H_0 = h\omega_1 a_f^+ a_f + h\omega_2 a_s^+ a_s + \varepsilon_1 S_{11} + \varepsilon_2 S_{22} + \varepsilon_3 S_{33}, \quad (1a)$$

$$H_1 = g_{xz} a_f^+ S_{xz} + g_{xz}^* a_f S_{xz} + g_{yz} a_s^+ S_{yz} + g_{yz}^* a_s S_{yz}. \quad (1b)$$

The first parts in $H_0$ of equation (1a) describes the propagating photon modes in the Sagnac waveguide loop with frequency $\omega_s$ and $\omega_f$, here $a_{s,f}^+$ and $a_{s,f}$ are creation and annihilation operators, respectively. The second parts in $H_0$ of equation (1a) describes the Hamiltonian of the atom, where $S_{ii}$ the number operator for the atom is $|i\rangle$ state and $\varepsilon_i$ is energy of this state. The interaction term ($H_1$) describes the scattering between photons and the atom, where $S_{\mu\nu} = b_\mu^+ b_\nu$ is the transition operators between the quantum state and $g_{\mu\upsilon}$ describe the interaction between each transition dipole moment and the photonic field. The transition operators between the quantum states satisfy the commutation relation $\left[S_{\mu\nu}, S_{\mu'\nu'}\right] = S_{\mu\nu'}\delta_{\mu'\nu} - S_{\mu'\nu}\delta_{\mu\nu'}$. We have different methods for the derivation of effective Hamiltonians. Quasi-degenerate perturbation theory [31, 32] is a method for the derivation of effective Hamiltonians in which the low-energy and high-energy subspaces are decoupled using a unitary transformation called the Schrieffer-Wolff transformation. We assume that the exact Hamiltonian is $H = H_0 + H_1 = H_0 + \zeta V$. Schrieffer-Wolff transformation is defined as $H_{eff} = e^{-S} H e^{S}$ where s is chosen such that the block of off-diagonal elements in $H_{eff}$ disappears up to the desired order. Using the Campbell-Baker-Hausdorff formula we can write $H_{eff}=e^{-s}He^{s}=\sum_{j=0}^{\infty}(1/j!)[H,s]^{(j)}$ . We can derive the equation $H_1+[H_0,s]=0$ and up to the second order in $\zeta$ the effective Hamiltonian yields

$$H_{eff}=H_0+\frac{1}{2}[H_1,s]. \tag{2}$$

If we consider

$$s=\alpha_{xz}g_{xz}a_f^+S_{xz}+\beta_{xz}g_{xz}^*a_fS_{zx}+\alpha_{yz}g_{yz}a_s^+S_{yz}+\beta_{yz}g_{yz}^*a_sS_{zy}, \tag{3}$$

we can obtain constants as: $\alpha_{xz}=-\beta_{xz}=1/(\varepsilon_z-\varepsilon_x-\hbar\omega_1)$ and $\alpha_{yz}=-\beta_{yz}=1/(\varepsilon_z-\varepsilon_y-\hbar\omega_2)$. Substituting relation (3) into (2), we can obtain

$$\begin{aligned}H_{eff}&=a_f^+a_f\left(\hbar\omega_f+(s_{zz}-s_{xx})|g_{xz}|^2\alpha_{xz}\right)+a_s^+a_s\left(\hbar\omega_s+(s_{zz}-s_{yy})\alpha_{yz}|g_{yz}|^2\right)+\\&+\varepsilon_xS_{xx}+\varepsilon_yS_{yy}+\varepsilon_zS_{zz}+|g_{xz}|^2\alpha_{xz}S_{zx}S_{xz}+|g_{yz}|^2\alpha_{yz}S_{zy}S_{yz}-\\&-\frac{1}{2}(\alpha_{xz}+\alpha_{yz})\left(S_{yx}g_{yz}g_{xz}^*a_s^{\dagger}a_f+S_{xy}g_{xz}g_{yz}^*a_f^{\dagger}a_s\right).\end{aligned} \tag{4}$$

We can describe the evolution of the system under consideration by the following wave function

$$\psi=c_1|1\rangle_s|0\rangle_f|y\rangle+c_2|0\rangle_s|1\rangle_f|x\rangle+c_3|0\rangle_s|0\rangle_f|z\rangle=c_1\psi_1+c_2\psi_2+c_3\psi_3. \tag{5}$$

Solving the Schrodinger equation $d\psi/dt=-(i/\hbar)H_{eff}\psi$, yields the governing equations of motion

$$\frac{dc_1}{dt}=-ic_1\left(\frac{\varepsilon_y}{\hbar}+\omega_s\right)+\frac{i}{2\hbar}(\alpha_{xz}+\alpha_{yz})g_{yz}g_{xz}^*c_2, \tag{6a}$$

$$\frac{dc_2}{dt}=-ic_2\left(\frac{\varepsilon_x}{\hbar}+\omega_f\right)-\frac{i}{2\hbar}(\alpha_{xz}+\alpha_{yz})g_{xz}g_{yz}^*c_1, \tag{6b}$$

$$\frac{dc_3}{dt}=-\frac{i}{\hbar}c_3\left(\varepsilon_z+|g_{xz}|^2\alpha_{xz}+|g_{yz}|^2\alpha_{yz}\right). \tag{6c}$$

Let us suppose that

$$\omega_a=\frac{\varepsilon_y}{\hbar}+\omega_s, \tag{7a}$$

$$\omega_b=\frac{\varepsilon_x}{\hbar}+\omega_f, \tag{7b}$$

$$\omega_c=\frac{1}{\hbar}\left(\varepsilon_z+|g_{xz}|^2\alpha_{xz}+|g_{yz}|^2\alpha_{yz}\right), \tag{7c}$$

$$\kappa = -\frac{g_{xz} g_{yz}^*}{2\hbar}(\alpha_{xz} + \alpha_{yz}) . \tag{7d}$$

After substituting equations (7a-7d) into equations of motion (6a-6c), the equations of motion reduce to

$$\frac{dc_1}{dt} = -i\omega_a c_1 - i\kappa c_2 , \tag{8a}$$

$$\frac{dc_2}{dt} = -i\omega_b c_2 - i\kappa^* c_1 , \tag{8b}$$

$$\frac{dc_3}{dt} = -ic_3\omega_h . \tag{8c}$$

From the solution of the Schrodinger equation, we obtain

$$c_1 = \exp\left[-\frac{i}{2}(\omega_a + \omega_b)t\right]\cos\left\{\left(\tfrac{1}{4}(\omega_a - \omega_b)^2 + |\kappa|^2\right)^{1/2} t\right\} - \\ -i\frac{\omega_a - \omega_b}{\left((\omega_a - \omega_b)^2 + 4|\kappa|^2\right)^{1/2}}\exp\left[-\tfrac{i}{2}(\omega_a + \omega_b)t\right]\sin\left\{\left(\tfrac{1}{4}(\omega_a - \omega_b)^2 + |\kappa|^2\right)^{1/2} t\right\}, \tag{9a}$$

$$c_2 = -\frac{i}{\kappa}\exp\left[-\frac{i}{2}(\omega_a + \omega_b)t\right]\frac{\omega_b^2 - \omega_a^2 + 2|\kappa|^2}{\left((\omega_a - \omega_b)^2 + 4|\kappa|^2\right)^{1/2}}\sin\left\{\left(\tfrac{1}{4}(\omega_a - \omega_b)^2 + |\kappa|^2\right)^{1/2} t\right\} . \tag{9b}$$

Choosing parameters in such a way that $\omega_a = \omega_b = \omega$, we get at real $\kappa$ the following equations

$$c_1 = e^{-i\omega t}\cos\{\kappa t\}, \tag{10a}$$

$$c_2 = -ie^{-i\omega t}\sin\{\kappa t\}. \tag{10b}$$

Equations (10a,10b) show that we have complete frequency conversion at $\kappa t = \pi/2$. So, we have operator of frequency transfer

$$Q_{FT}|10\rangle = |01\rangle \tag{11}$$

at these conditions.

## 4. Implementing quantum NOT and CNOT gates

Let's introduce logical qubit encoded by one photon that can be situated in two states with frequencies $\omega_{s_1}$ and $\omega_{s_2}$. We consider a three level atom excited to level $|y\rangle$ by an arbitrary photon logical qubit at Raman transition

$$|\psi_1\rangle_L = \alpha|1\rangle_{s1}|0\rangle_{s2}|0\rangle_f|y\rangle + \beta e^{i\phi}|0\rangle_{s1}|1\rangle_{s2}|0\rangle_f|y\rangle \equiv \alpha|0\rangle_L + \beta e^{i\phi}|1\rangle_L . \tag{12}$$

By implementing frequency conversion $\omega_{s1} \to \omega_f$ on the above state, we obtain:

$$|\psi_2\rangle_L = \hat{Q}_{FT}(1,f)|\psi_1\rangle_L = \alpha|0\rangle_{s1}|0\rangle_{s2}|1\rangle_f|x\rangle + \beta e^{i\phi}|0\rangle_{s1}|1\rangle_{s2}|0\rangle_f|y\rangle . \tag{13}$$

And acting by quantum operator once again, $\omega_{s2} \leftrightarrow \omega_f$, we obtain the wave function

$$|\psi_3\rangle_L = \hat{Q}_{FT}(2,f)|\psi_2\rangle_L = \alpha|0\rangle_{s1}|1\rangle_{s2}|0\rangle_f|y\rangle + \beta e^{i\phi}|0\rangle_{s1}|0\rangle_{s2}|1\rangle_f|x\rangle. \quad (14)$$

And at last converting: $\omega_{s1} \leftarrow \omega_f$,

$$\hat{Q}_{FT}(1,f)|\psi_3\rangle_L = \alpha|0\rangle_{s1}|1\rangle_{s2}|0\rangle_f|y\rangle + \beta e^{i\phi}|1\rangle_{s1}|0\rangle_{s2}|0\rangle_f|y\rangle = NOT|\psi_1\rangle_L. \quad (15)$$

After each step, photon logical is stored in quantum memory. During storage time the position of level $|y\rangle$ is changed by the Stark effect. Then logical qubit is extracted from the processor unit and directed again to processing unit.

Quantum logic gates perform manipulations of the information, converting it from one form to another. It is worth noting that quantum gates are positioned in the heart of quantum computing. We exploit the idea of photon logical qubits where single photon excitation is distributed among two photon states with various frequencies. We have used a phenomenological Hamiltonian to describe the evolution of the photon logical qubit where atomic variables are accounted as classical oscillators. With that, we consider the evolution of the system beginning from sublevel two of three level atom ground state. In the case of NOT operation this sublevel is populated deterministically via Raman process using two classical pulses. Then NOT operation is accomplished in several steps connected with transformation of the photon with frequency of logical qubit to an auxiliary photon and back. The choice of necessary photon frequency is achieved by shifting atomic levels using the Stark effect.

By implementing several times the quantum operator $\hat{Q}_{FT}$, we arrive at the operator NOT. In the case of CNOT operation we take for the population of level $|y\rangle$ one classical pulse and one photon from control qubit. Therefore, NOT operation proceeds when the photon of the control qubit is present. If there is no photon with necessary frequency in the state of the control qubit, NOT operation does not proceed. So, we have CNOT operation in whole.

**Acknowledgments**
This research was supported by the Russian Foundation for basic research (Grant No. 15-02-05478).

**References**
[1] Huang J M and Kumar P 1992 *Phys. Rev. Lett.* **68** 2153
[2] Tanzilli S, Tittel W, Halder M, Alibart O, Baldi P, Gisin N and Zbinden H 2005 *Nature* **437** 116
[3] Ikuta R, Kusaka Y, Kitano T, Kato H, Yamamoto T, Koashi M and Imoto N 2011 *Nat. Commun.* **2** 537
[4] Zaske S *et al* 2012 *Phys. Rev. Lett.* **109** 147404
[5] Porat G, Silberberg Y, Arie A and Suchowski H 2012 *Opt. Express* **4** 3613
[6] Agha I, Davanço M, Thurston B and Srinivasan K 2012 *Opt. Lett.* **37** 2997
[7] Wang G, Xue Y, Wu J H, Kang Z H, Jiang Y, Liu S S and Gao J Y 2010 *Opt. Lett.* **35** 3778
[8] Jen H H and Kennedy T A B 2010 *Phys. Rev.* A **82** 023815
[9] Dong C, Fiore V, Kuzyk M C and Wang H 2012 *Science* **338** 1609
[10] Hill J T, Safavi-Naeini A H, Chan J and Painter O 2012 *Nat. Commun.* **3** 1196
[11] Preble S, Cao L, Elshaari A, Aboketaf A and Adams D 2012 *Appl. Phys. Lett.* **101** 171110
[12] Bradford M, Obi K C and Shen J T 2012 *Phys. Rev. Lett.* **108** 103902
[13] Bradford M and Shen J T 2012 *Phys. Rev.* A **85** 043814
[14] Wu Y and Deng L 2004 *Opt. Lett.* **29** 1144

[15] Payne M G and Deng L 2003 *Phys. Rev. Lett.* **91** 123602
[16] Wu Y and Yang X X 2004 *Phys. Rev.* A **70** 053818
[17] Wu Y, Payne M G, Hagley E W and Deng L 2004 *Phys. Rev.* A **69** 063803
[18] Wang Z H, Zhou L, Li Y and Sun C P 2014 *Phys. Rev.* A **89** 053813
[19] Li G X, Tan H T and Macovei M 2007 *Phys. Rev.* A **76** 053827
[20] Albota M A and Wong F C 2004 *Opt. Lett.* **29** 1449-51
[21] Langrock C, Diamanti E, Roussev R V, Yamamoto Y, Fejer M M and Takesue H 2005 *Opt. Lett.* **30** 1725-28
[22] Vandevender A P and Kwiat P G 2004 *J. Mod. Opt.* **51** 1433
[23] Ding D S, Zhou Z Y, Huang W, Shi B S, Zou X B and Guo G C 2012 *Phys. Rev.* A **86** 033803
[24] Gibbs H M 1985 *Optical Bistability: Controlling Light with Light* (Orlando: Academic)
[25] Takesue H 2010 Single-photon frequency down-conversion experiment *Phys. Rev.* A **82** 013833
[26] Radnaev A G *et al* 2010 A quantum memory with telecom-wavelength conversion *Nature Phys.* **6** 894
[27] De Greve K *et al* 2012 *Nature* **491** 421
[28] Zheng A et al 2014 *J. Phys.* B **47** 055501
[29] Yan W-B, Huang J-F and Fan H 2013 *Sci. Rep.* **3** 3555
[30] Zheng A *et al* 2015 *Laser Phys.* **25** 065201
[31] Roland Winkler 2003 Spin-Orbit Coupling Effects in Two-Dimensional Electron and Hole System *Springer Tracts in Modern Physics* **191** (Berlin Heidelberg: Springer Verlag)
[32] Bravyi S, Divincenzo D P and Loss D 2011 *Annals of Physics* **326** 2793–2826

Journal of Physics: Conference Series **714** (2016) 012003 doi:10.1088/1742-6596/714/1/012003

# Investigations of properties of opal-like photonic crystals using combined methods

**A A Akhmadeev[1] and M Kh Salakhov[1,2]**

[1] Institute of Physics, Kazan Federal University, 420008, 18 Kremlevskaya St., Kazan, Russian Federation

[2] Tatarstan Academy of Sciences, 420111, 20 Bauman St., Kazan, Russian Federation

E-mail: Albert.Akhmadeev@kpfu.ru

**Abstract.** We propose a technique that combines atomic force microscopy, optical spectrophotometry and diffraction method for characterization of photonic crystals. These materials are promising for the creation of various devices, as well as for the observation of new effects due to their unique properties. Photonic crystals were synthesized by self-assembly of colloidal silica particles with low deviation in size. It is shown that the developed technique allows to increase the accuracy of determining the parameters of the photonic crystal.

## 1. Introduction

A photonic crystal (PC) is a material with a periodic spatial distribution of the refractive index having a photonic bandgap in the spectrum of electromagnetic states [1]. A photonic crystal is a promising material for creation of various devices such as low-threshold lasers, high-efficiency light-emitting diodes, waveguides of a new type, optical switches and filters, and ultra-fast optical computers in the future. At the moment there are numerous approaches to the creation of photonic crystals based on lithography [2], interference holography [3] and self-assembly of colloidal particles [4]. In particular, a method based on self-assembly of colloidal particles with low deviation in size is cheap and simple to implement and allows to obtain a large number of samples of artificial opals for a short time. The samples produced by the majority of existing methods have various structural defects, which significantly affect their optical properties. Thus, our task is the improvement of methods for determining the characteristics of the photonic crystals. In addition, it could be crucial for the manifestation of a new quantumelectrodynamical effect predicted in earlier works [5]. In this study the photonic crystals were synthesized by self-assembly of colloidal particles of silicon dioxide [6]. The parameters of the photonic crystals, such as the lattice period, particle size and effective refraction index were investigated by complementary techniques, including atomic force microscopy (AFM), optical spectrophotometry and the diffraction method. It is shown that the measurement accuracy increases with the combined use of these techniques.

## 2. Spectroscopic method

Period of the PC structure, particle size and the effective refractive index can be determined by transmission spectra measured at different incidence angles. Taking account of Snell's law, the Bragg's law is written as:

$$\lambda = 2d_{(111)}\sqrt{n_{eff}^2 - \sin^2\theta}, \tag{1}$$

where $\lambda$ is the wavelength corresponding to the bandgap, $d_{(111)}$ is the period of the structure, i.e. the distance between the (111) planes, $n_{eff}$ is the effective refractive index, $\theta$ is the angle between the normal to the sample surface and the direction of the incident light. The effective refractive index $n_{eff}$ of the photonic crystal depends on the refractive index of the $SiO_2$ particles ($n_{SiO_2} = 1.45$), the refractive index of air ($n_{air} = 1$) and the volume fraction of microspheres (for an ideal close-packed structure of spheres $f = 0.7405$):

$$n_{eff} = \sqrt{n_{SiO_2}^2 \cdot f + n_{air}^2 \cdot (1 - f)}. \tag{2}$$

Figure 1 (a) shows the transmission spectra of PC samples based on silica. The dependence of band gap spectral position ($\lambda$) on light incidence angle ($\theta$) is linear in terms of $\lambda^2 - \sin^2\theta$ (figure 1 (b)).

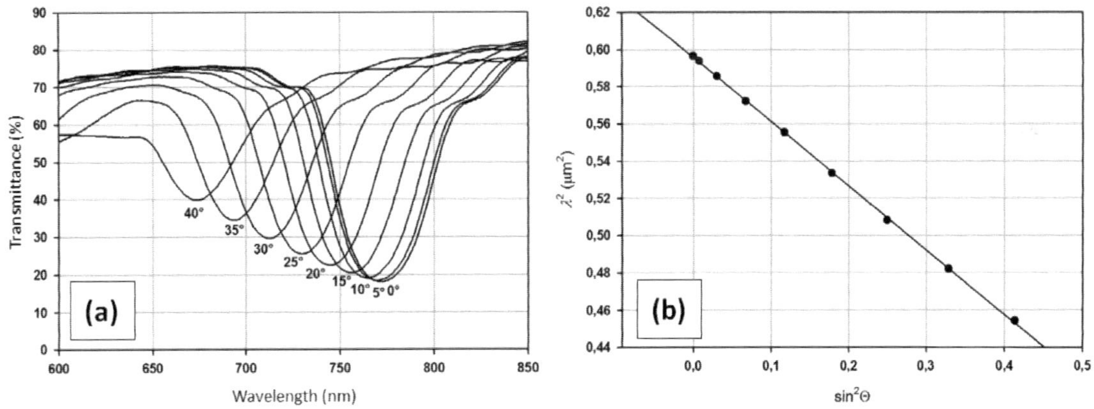


**Figure 1.** a) Transmission spectra of the PC sample. b) The spectral position of the band gap for different angles of incidence of light on the sample surface.

Using a linear least-squares approximation, we have calculated the effective refractive index ($n_eff = 1.31 \pm 0.07$) as well as the period of the structure ($d_{(111)} = 388 \pm 19$ nm) and the particle size ($D = 475 \pm 23$ nm). In this case the particle size is determined by the period as follows:

$$D = \sqrt{\frac{3}{2}}d. \tag{3}$$

## 3. Diffraction on a periodic structure

In the previous section, the occurrence of photonic bandgaps (the dips in the transmission spectra) are caused by diffraction and interference of the transmitted light of a certain wavelength on the periodically alternating layers (parallel to the sample surface), consisting of close-packed spherical particles of silicon dioxide. In turn, the PC, which are grown by self-assembly of the microparticles have a periodicity in three dimensions (three-dimensional photonic crystal). Each layer of the microparticles in such crystals can be considered as a 2D diffraction grating with hexagonal symmetry (see figure 2). The positions of diffraction maxima satisfy the equation:

$$d\sin\theta = k\lambda, \tag{4}$$

where $d$ is the lattice period (the lateral period of PC structure), $\theta$ is the angle of diffraction, $k$ is the order of diffraction, $\lambda$ is the wavelength of the light.

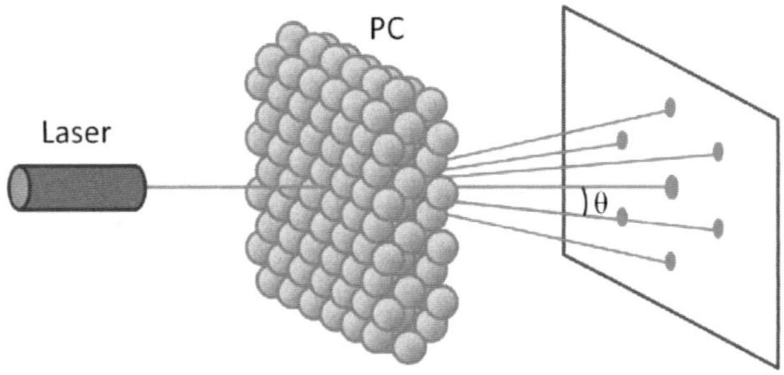


**Figure 2.** Scheme for laser diffraction experiment. The laser wavelength is 532 nm.

The diffraction pattern allows you to quickly and easily determine the average size of the microparticles of the photonic crystal, taking into account the fact that the diameter of the spheres $D$ is associated with the period lattice $d$ by the following equation:

$$D = \frac{2}{\sqrt{3}} d. \tag{5}$$

The laser diffraction method is not applicable when $d < \lambda$. Using this method the diameter of spheres was measured, $D = 841 \pm 46$ nm, as confirmed by atomic force microscopy.

## 4. Combined method

Atomic force microscopy, optical spectroscopy and the laser diffraction method complement each other. By combining these methods it is possible to improve significantly the accuracy of the measurement of the average particle size. The idea of a joint consideration of the measurement error of the same parameter using different methods is based on the statistical nature of the measurement error. Assume that the parameter $x$ measured by the method $A$ is normally distributed:

$$\rho_A(x) \sim \exp\left(-\frac{(x - x_A)^2}{2\sigma_A^2}\right), \tag{6}$$

with its own mean value $x_A$ and standard deviation $\sigma_A$. If we additionally measure the same parameter using method $B$ with a mean value $x_B$ and standard deviation $\sigma_B$, the probability density according to the probability theory will be a multiplication $\rho_A(x)\ \ \rho_B(x)$:

$$\rho(x) \sim \exp\left(-\frac{(x - x_B)^2}{2\sigma_B^2} - \frac{(x - x_B)^2}{2\sigma_B^2}\right) = F\exp\left(-\frac{(x - x_C)^2}{2\sigma_C^2}\right), \tag{7}$$

where $x_C = (\sigma_B^2 x_A + \sigma_A^2 x_B)(\sigma_A^2 + \sigma_B^2)^{-1}$ and $\sigma_C = \sigma_A \sigma_B (\sigma_A^2 + \sigma_B^2)^{-\frac{1}{2}}$ are the mean and standard deviation of the resulting distributions, $F = \exp\left(-\frac{(x_A - x_B)^2}{2(\sigma_A^2 + \sigma_B^2)}\right)$. Thus, the combined use of several methods for measuring the same parameter can reduce the experimental error due to the fact that the value of $\sigma_C$ is less than $\sigma_A$ and $\sigma_B$. This approach is used to determine the proton's charge radius by combining the methods of precision laser spectroscopy and electron-proton scattering [7]. In addition, the same approach is applied in optical spectroscopy in the determination of the proton radius by different transitions [8].

Applying this approach to measure the particle size of the same sample using the methods of atomic force microscopy, optical spectroscopy and diffraction method, it was possible to increase the accuracy of determining the average size of silica particles (see figure 3). Detection of particles on AFM images and determination of their size were performed using the method described in [9].

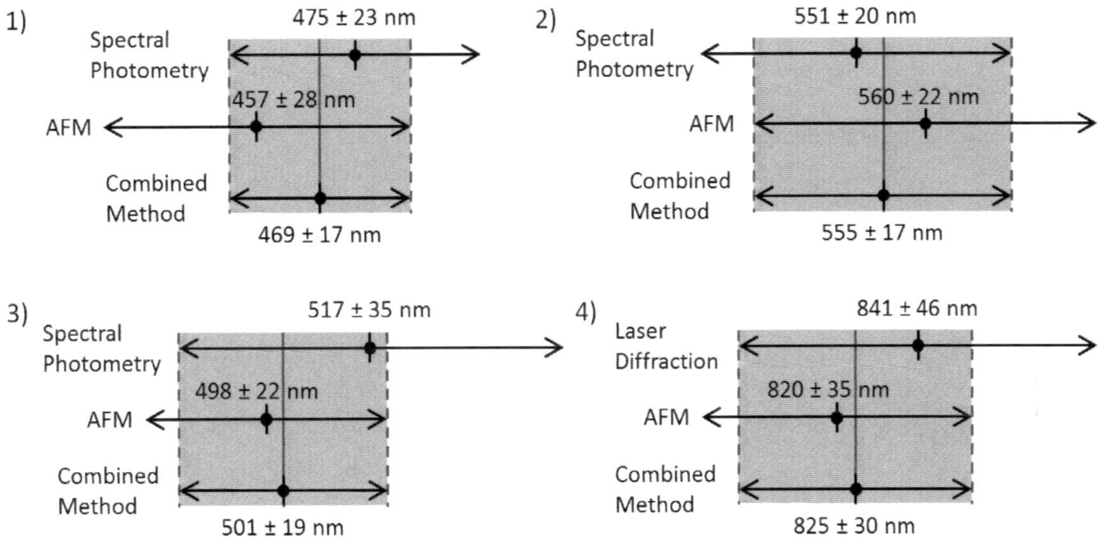


**Figure 3.** The results of measuring the average particle diameter by combining different methods (four PC samples).

## 5. Conclusion

We have developed a method that combines spectral photometry, atomic force microscopy and laser diffraction for characterization of the synthesized samples of photonic crystals, allowing to measure the diameter of the microspheres, the effective refractive index and the average size dispersion. The developed method allows to increase the accuracy of determining the average particle size of the photonic crystal.

## Acknowledgments

This work was funded by the subsidy allocated to Kazan federal University for the state assignment in the sphere of scientific activities.

## References

[1] Yablonovitch E 1987 *Phys. Rev. Lett.* **58** 2059–2062
[2] Yao P, Schneider G J, Prather D W, Wetzel E D and O'Brien D J 2005 *Optics Express* **13**(7) 2370–2376
[3] Duneau M, Delyon F and Audier M 2004 *J. Appl. Phys.* **96**(5) 2428–36
[4] Vlasov Y A, Bo X-Z, Sturm J C and Norris D J 2001 *Nature* **414** 289–293
[5] Gainutdinov R K, Khamadeev M A and Salakhov M K 2012 *Phys. Rev.* A **85** 0538362
[6] Stober W, Fink A and Bohn E 1968 *J. Colloid Interface Sci.* **26** 62–69
[7] Mohr P J, Taylor B N and Newell D B 2012 *Rev. Mod. Phys.* **84** 1527–1605
[8] Pohl R, Gilman R, Miller G A and Pachucki K 2013 *Ann. Rev. Nucl. Part. Sci.* **63** 175–204
[9] Akhmadeev A A and Salakhov M Kh 2014 *J. Phys.: Conf. Ser.* **560** 012005

# A simple way of introducing stochastic differential equations

**A M Basharov**

Moscow Institute of Physics and Technology, 141700 Dolgoprudny, Russia
National Research Centre «Kurchatov Institute», 123182 Moscow, Russia

E-mail: basharov@gmail.com; http://basharov.me

**Abstract.** The notion of the Ito increment and the stochastic differential equation of the non-Wiener type were introduced using the simple "natural" property of counting process. The properties of the stochastic differential and integral were demonstrated and clarified in a simple and original way.

## 1. Introduction. SDE paradigm

One of the primary methods for the world knowledge is the research into the system response to small (and not small) perturbations, in particular, the study of the system return to equilibrium or quasi-equilibrium state after its removal from one of this or other condition of the kind. These are known as relaxation processes, and many problems of physics affect the relaxation of various kinds of excitation in the atomic, photonic, phonon and other systems. Similar processes can be found not only in physical, chemical, biological and social systems, but also in such seemingly distant sections as models of financial markets and others.

In the study of the system response to small (and not small) perturbations and/or parameter variations virtually all familiar concepts keep emerging. Indeed, first it is necessary to parameterize the system under study and to identify independent and dependent parameters, such as $y = y(x)$. Then we consider their variations or increments $\Delta y$ and $\Delta x$. In the case of small increments, they appear to be related by direct proportional dependence

$$\Delta y(x) \equiv y(x+\Delta x) - y(x) \cong C_1(x)\Delta x ,$$

and the value $C_1(x)$ is the characteristics of the studied system which can be defined experimentally and/or which is obedient to some laws. So, in fact, there arises a concept of derivative since $C_1(x) = y'(x)$ for a «good» function $y(x)$. As a result, the basic laws are expressed by differential equations. For example, the position of the material point of mass $m$ at every point of time $t$ on the axis $x$ is determined as the coordinate $x$. Therefore we have the parameter dependence $x = x(t)$. Parameter increments define concepts such as velocity $v_x$ and acceleration $a_x$ (projections of velocity and acceleration on the axis $x$), $\Delta x(t) \cong v_x(t)\Delta t$, $\Delta v_x(t) \cong a_x(t)\Delta t$, and acceleration appears to be related to forces acting on the particle by Newton's second law of motion

$$ma_x = \sum F_x .$$

Assuming that among the forces acting on the particle there can be either deterministic $F$ or random forces $X(t)$, we obtain the differential equation with random values

$$m\ddot{x}(t) = F + X(t)\,. \tag{1}$$

To characterize a random force, we note that when a particle (material point) moves across the environment there arise occasional collisions with environment particles. Let all the environment particles be divided into groups numbered by the index $i$, such groups that for each collision of the material point concerned with the particles of the $i$-th group, a material point obtains a value increment of momentum $\delta_k$. Then it is convenient to rewrite the Newton's equation with a random force in the form of

$$\Delta p_x(t) = F\Delta t + \sum_k \delta_k \Delta N(\delta_k, t)\,,$$

where $p_x$ is the momentum of a material point, and $\Delta N(\delta_k, t)$ is the number of collisions of a material point with the $k$-th group particles of environment per time $\Delta t$. Value $\Delta N(\delta_k, t)$ is the increment of some integer random value $N(\delta_k, t)$: $\Delta N(t) = N(\delta_k, t + \Delta t) - N(\delta_k, t)$.

Another example is provided by any electrical circuit. For example, if we consider the definition of electric current, then the charge $\Delta Q$ passing over a time $\Delta t$ across the specified section of the conductor is $\Delta Q(t) = I(t)\Delta t$, where $I(t)$ is the electric current intensity. However, given that the charge is transferred by the elementary carriers of charge $q_0$, then

$$\Delta Q(t) = q_0 \Delta N(t)\,, \tag{2}$$

where $\Delta N(t)$ is the number of elementary charge carriers passing over a time $\Delta t$ across the specified section of the conductor.

If the flow of charged particles is quite rare, for example, when driving the elementary charge carriers in electron tubes and the current pulse caused by the charge hitting the anode is very short (small inductance), the equation for the current can be conveniently written as

$$\Delta I(t) = -\alpha I(t)\Delta t + \beta \Delta N(t)\,, \tag{3}$$

where $\Delta N(t) = \sum_i \Delta N_i(t)$. If the time points of charge hitting the anode are designated as values $t_k$, the equation for the current can also be written as

$$\dot{I}(t) = -\alpha I(t) + \beta \sum_k \delta(t - t_k)\,. \tag{4}$$

The constant $\alpha$ is related to the constant $\beta$ by the ratio $-\alpha q_0 + \beta = 0$ since before and some time after the charge hitting the anode, the current caused by one elementary charge should be equal to zero

$$\sum \Delta I(t) = -\alpha q_0 + \beta\,.$$

The ratio $-\alpha q_0 + \beta = 0$ remains valid for a more intense flux of electrons if the electrons enter the anode and affect it independently. As an example of equations (1)-(4) we will consider the features of stochastic differential equations (SDEs) at the elementary level and identify the basic concepts arising in this case.

## 2. Peculiarities of SDEs and stochastic integrals. The Ito increment algebra

We will discuss main differences of random terms of SDEs and stochastic integrals from the ordinary differential equations and the Riemann integrals

In the examples given above a random term was expressed as some value $N(t)$ whose increment

$$\Delta N(t) = N(t+\Delta t) - N(t)$$

provided for an integer number of random events happening in a time interval $\Delta t$. At the same time, it is evident that the main incremental rule is fulfilled – the sum of increments during shorter successive intervals is equal to the increment during the entire interval:

$$\Delta N(t) = \sum_{i=1}^{M} \Delta N_i = N(t+\Delta t) - N(t).$$

Here the time interval $[t, t+\Delta t]$, $\Delta t > 0$ is divided by points $t = t_0 < t_1 < t_2 < \ldots < t_M = t + \Delta t$ into $M$ small subintervals $\Delta t_i = t_i - t_{i-1}$ и $\Delta N_i = N(t_i) - N(t_{i-1})$. Thus, in partitioning the time interval into shorter small subintervals, it is possible to reach the situation when $\Delta N_i$ take only two values 0 and 1, which describes only two feasible options – whether some event has happened or not.

Thus, when choosing a sufficiently short interval of time $\Delta t_i$, we have an exact equality $\Delta N_i^2 = \Delta N_i$, because for shorter time intervals $\Delta N_i$ can take only the values 0 and 1, and $0^2 = 0$ and $1^2 = 1$. Such small time intervals and the corresponding increments of random process $N(t)$ are to be discussed as Ito increments $dt$ and $dN(t)$, respectively. Also, the terms differential or the Ito differential are often used. However, unlike the conventional analysis by which the differential $df(x)$ of function $f(x)$ indicates only that it is a linear (in increment of the argument) part of the increment of the function

$$\Delta f(x) = df(x) + C_2(x)(\Delta x)^2 + C_3(x)(\Delta x)^3 + \ldots, \tag{5}$$

the Ito differential

$$dN(t) = N(t+dt) - N(t) \tag{6}$$

suggests that the time interval $dt$ is that during this time only one event happens or does not happen, such as an electron hitting the anode. The possibility of simultaneous entrance of two electrons to the anode is neglected.

So, the random process $N(t)$ consisting in counting the number of the events happened over the time $t$, otherwise the reading process, takes integer values, with increment (6) of the random process satisfying algebra

$$dN(t)dN(t) = dN(t). \tag{7}$$

This algebra determines the main difference between SDEs and integrals containing a random process $N(t)$ and the conventional differential and integral calculus.

Recall that in terms of the function increment, the main point determining an ordinary differential and integral calculus is the fact that for small increments of the argument $|\Delta x| << 1$ we have the following relation

$$|\Delta x| >> |\Delta x|^2 >> |\Delta x|^3 >> \ldots,$$

so in the expansion of the increment $\Delta f(x)$ of a «good» function $f(x)$,

$$\Delta f(x) = C_1(x)\Delta x + C_2(x)(\Delta x)^2 + \ldots$$

at $|\Delta x| << 1$ only $\Delta x$ linear term $\Delta f(x) \approx df(x)$ matters, which is referred to as differential regardless of the value of the argument increment $\Delta x$. Obviously, in the case of dependence of the "good" function $f$ on $N(t)$ all the terms should be taken into account in increment $\Delta f$

$$\begin{aligned}\Delta f(N(t)) &= f(N(t)+\Delta N(t)) - f(N(t) = \tfrac{1}{1!}f'(N(t))\Delta N(t) + \tfrac{1}{2!}f''(N(t))(\Delta N(t))^2 + \ldots = \\ &= [\tfrac{1}{1!}f'(N(t)) + \tfrac{1}{2!}f''(N(t)) + \ldots]\Delta N(t) = [f(N(t)+1) - f(N(t))]\Delta N(t).\end{aligned} \quad (8)$$

Otherwise, while using increments

$$df(N(t)) = f(N(t)+dN(t)) - f(N(t) = [f(N(t)+1) - f(N(t))]dN(t). \quad (9)$$

This difference can be seen from the formula for differentiating a function product. For example,

$$\begin{aligned}dN^2(t) &= N^2(t+dt) - N^2(t) = (N(t)+dN(t))^2 - N^2(t) = 2N(t)dN(t) + (dN(t))^2 = \\ &= 2N(t)dN(t) + dN(t) = (2N(t)+1)dN(t),\end{aligned} \quad (10)$$

i.e. in contrast to the ordinary Leibniz formula, in differentiating the function product it is necessary to take into account increment products if these functions are random processes, or functions of random processes.

Another feature of random function $N(t)$ occurs in the concept of a stochastic integral, i.e. an integral over increments of a stochastic process. Let us discuss how to understand the integral

$$\int_{t_0}^{t} N(t')dN(t'). \quad (11)$$

It is natural to represent it as the limit of integral sums $\int_{t_0}^{t} N(t')dN(t') = \lim_{\substack{M\to\infty \\ \max \Delta t_i \to 0}} \sum_{i=1}^{M} N(t_i')\Delta N_i$, where time interval $[t_0, t]$, $t > t_0$ is divided by points $t = t_0 < t_1 < t_2 < \ldots < t_M = t$ into $M$ small subintervals, $\Delta N_i = N(t_i) - N(t_{i-1})$, and points $t_i' \in [t_{i-1}, t_i]$.

As is known from the probability theory there are several concepts of limit for random values. We will note a curious fact that is independent of the concept of limit for random values. We will discuss how point $t_i'$ within the partition interval should be chosen. In calculating the integrand at the left-edge of partition intervals

$$t_i' = t_{i-1},\ i = 1, \ldots, M,$$

then the resulting limit to integral sums will be designated as

$$S_L = \lim_{\substack{M\to\infty \\ \max \Delta t_i \to 0}} \sum_{i=1}^{M} N(t_{i-1})\Delta N_i.$$

It is not hard to understand that this limit will be different from the limit of Riemann sums, when the integrand is evaluated at the right edge of partition intervals

$$S_R = \lim_{\substack{M\to\infty \\ \max \Delta t_i \to 0}} \sum_{i=1}^{M} N(t_i)\Delta N_i = \lim_{\substack{M\to\infty \\ \max \Delta t_i \to 0}} \sum_{i=1}^{M} (N(t_{i-1}) + \Delta N_i)\Delta N_i = S_L + \lim_{\substack{M\to\infty \\ \max \Delta t_i \to 0}} \sum_{i=1}^{M} \Delta N_i = S_L + N(t) - N(t_0).$$

Stochastic integrals defined by the value of the integrand at the left-edge of partition intervals, with account of the relevant definition of the limit to integral comprise a class of the so-called stochastic Ito integrals. The choice of the left-edge of partition interval is useful for the so-called non-anticipating functions when a would-be behaviour of a random process is statistically independent of its current state. For such functions $F(t)$ the mean value of the stochastic Ito integral has a simple representation

$$<\int_{t_0}^{t} F(t')dN(t')>=\int_{t_0}^{t}<F(t')><dN(t')>.$$

Selecting the Ito integral as a stochastic integral leads to the unconventional formulas of integration, with the differentiation formulas being consistent with integration, such as the value of (11) can be derived from (10) as

$$\int_{t_0}^{t} N(t')dN(t')=\tfrac{1}{2}\int_{t_0}^{t} dN^2(t)-\tfrac{1}{2}\int_{t_0}^{t} dN(t')=\tfrac{1}{2}(N^2(t)-N^2(t_0))-\tfrac{1}{2}(N(t)-N(t_0)).$$

On the other hand, considering the integral as a limit of Riemann sums with values of the integrand at the left-edge of partition intervals, we have

$$\int_{t_0}^{t} N(t')dN(t')=\lim_{\substack{M\to\infty \\ \max\Delta t_i\to 0}}\sum_{i=1}^{M} N(t_{i-1})\Delta N_i=\frac{1}{2}\lim_{\substack{M\to\infty \\ \max\Delta t_i\to 0}}\sum_{i=1}^{M}[(N(t_{i-1})+\Delta N_i)^2-N^2(t_{i-1})-(\Delta N_i)^2]=$$

$$=\frac{1}{2}\lim_{\substack{M\to\infty \\ \max\Delta t_i\to 0}}\sum_{i=1}^{M}[(N(t_{i+1}))^2-N^2(t_{i-1})-(\Delta N_i)^2]=\frac{1}{2}\lim_{\substack{M\to\infty \\ \max\Delta t_i\to 0}}[N^2(t)-N^2(t_0)-(N(t)-N(t_0))]=$$

$$=\frac{1}{2}[N^2(t)-N^2(t_0)-(N(t)-N(t_0))].$$

This result does not depend on the concept of limit used. It demonstrates the consistency of algebraic relations (7) with the concept of stochastic Ito integral. Thus, stochastic differential equations and stochastic integrals (and related stochastic integral equations) are somewhat "two sides of the same coin". And the concept of stochastic continuity will allow separating random functions from discontinuous deterministic functions, which can demonstrate the specified features at certain partitions.

Without going too deep into the theory, it is natural to hypothesize the nature of the averaging of increment $dN(t)$. Being a good idealization, random process $N(t)$ should be in some sense homogeneous with the average value $dN(t)$ proportional to the time interval

$$<dN(t)>=\lambda dt. \tag{12}$$

Parameter $\lambda$ is to be referred to as velocity or intensity of a random process $N(t)$.

Finally, if you give a standard definition for a limit for the partial sums of stochastic integral as the limit in the mean square, it is easy to prove that the integrals of non-anticipating functions take the following form $\int_{t_0}^{t} F(t')dN(t')dt'=0$, which allows adding an important ratio to the increment algebra (7) and (12)

$$dN(t)dt=0. \tag{13}$$

Moreover, the requirement for statistical independence $dN(t)$ and $dN(t')$ in the case of non-overlapping time intervals $dt$ and $dt'$ leads to

$$<dN(t)dN(t')>=<dN(t)><dN(t')>,\ t\neq t'. \tag{14}$$

Finally, SDEs (4) can be written in terms of the increment of the random process $N(t)$ (with algebra increments (7) (12) - (14)) in the form

$$dI(t) = -\alpha I(t)dt + \beta dN(t). \tag{15}$$

In terms of the algebra increments (7) (12) - (14) ordinary differential calculus is characterized by the algebra

$$dtdt = 0,$$

if value $dt$ is meant to be a rather small increment, i.e. the Ito increment. Then the formula for the Ito increment of function $f(t)$ takes on the following meaning

$$df(t) = f(t+dt) - f(t) = f'(t)dt + \tfrac{1}{2!}f''(t)(dt)^2 + ... = f'(t)dt.$$

## 3. SDEs and kinetic equations

Equations (7) and (12)-(14) determine the so-called Poisson process that essentially defines the entire apparatus of the SDEs. Let us see how the SDEs are linked to kinetic equations of the system that underlie the dynamics of open systems.

Let $f(t)$ be a continuously differential function, and $P(n,t)$ is the probability of the fact that by the time $t$ random value $N(t)$ has the meaning $n$. Then it is possible to introduce the average value $<f(N(t))>$ in the form $<f(N(t))>=\sum_{n=0}^{\infty} f(n)P(n,t)$. Let us consider equation (9): $df(N(t)) = dN(t)\{f(N(t)+1) - f(N(t))\}$ and average it by means of $P(n,t)$,

$$d<f(N(t))>=\lambda dt<f(N(t)+1)-f(N(t))>, \quad \frac{d<f(N(t))>}{dt} = \lambda <f(N(t)+1)-f(N(t))>,$$

$$\sum_{n=0}^{\infty} f(n)\frac{\partial P(n,t)}{\partial t} = \lambda\sum_{n=0}^{\infty}\{f(n+1)P(n,t) - f(n)P(n,t)\} = \lambda\sum_{n=1}^{\infty} f(n)P(n-1,t) - \lambda\sum_{n=0}^{\infty} f(n)P(n,t).$$

Supposing $f(t)$ to be an arbitrary function with $f(0)=0$, we derive the equation

$$\frac{\partial P(n,t)}{\partial t} = \lambda[P(n-1,t) - P(n,t)]. \tag{16}$$

This equation is an example of the kinetic equation derived from the SDE.

The solution of equation (16) by means of the generative function gives rise to the so-called Poisson distribution

$$P(n,t) = e^{-\lambda t}\frac{\lambda^n t^n}{n!}, \tag{17}$$

and value $N(t)$ is also called the Poisson process.

The average value of the number of electrons at the anode $\bar{n}$ and the variance of this number $\sigma^2 = <n^2> - \bar{n}^2$ for the Poisson distribution prove to be equal to each other. These results are not difficult to obtain from differential algebra (7) and (12) - (14). Since

$$N(t) = \int_0^t dN(t'),$$

calculation of the average value and the variance of value $N(t)$ can be presented in the following way:

$$< N(t) >= \int_0^t < dN(\tilde{t}) >= \int_0^t \lambda d\tilde{t} = \lambda t ,$$

$$< [N(t) - \lambda t]^2 >=< [N(t)]^2 > -2\lambda t < N(t) > +(\lambda t)^2 =< [N(t)]^2 > -(\lambda t)^2 ,$$

$$< [N(t)]^2 >=< \int_0^t dN(\tilde{t}) \int_0^t dN(\tilde{\tilde{t}}) >= \lim < \sum_i \Delta N_i \sum_j \Delta N_j >= \lim\{< \sum_i (\Delta N_i)^2 > +2\sum_{i>j} < \Delta N_i \Delta N_j >\}$$
$$= \lim\{< \sum_i \Delta N_i > +2\lambda^2 \sum_{i>j} \Delta t_i \Delta t_j\} = \lambda t + \lim\{\lambda^2 \sum_i (\Delta t_i)^2 + 2\lambda^2 \sum_{i>j} \Delta t_i \Delta t_j - \lambda^2 \sum_i (\Delta t_i)^2\} = \lambda t + (\lambda t)^2.$$

## 4. Conclusion

The proposed approach to introducing SDEs differs from the conventional ones (see, for example, [1-3]): one is immediately given an idea of non-Wiener type SDEs, which serve the basis for standard Wiener type SDEs derived by means of the central limit theorem [4]. Universality of stochastic processes underlying the SDEs is due to the central limit theorem and different time scales for deterministic and random terms. Their kinetic equations may contain fractional derivatives with respect to spatial variables, reflecting the self-similarity and fractal properties of the medium [4,5]. Moreover, the very random processes are the Markov processes. According to the approach considered above, non-Markov random processes with memory effect arise as the subordinated processes, and their kinetic equations contain terms with fractional time derivative [5]. However, the straightforward derivation of their kinetic equations similar to the ones considered in [4] for Markov processes is unknown to the author.

## References

[1] Jacobs K 2010 *Stochastic Processes for Physicists* (Cambridge: CUP)
[2] Breuer H-P and Petruccione F 2002 *Theory of open quantum systems* (Oxford: OUP)
[3] Cont R and Tankov P 2004 *Financial modeling with jump processes* (London: CRC Press)
[4] Basharov A M 2013 *J. Phys.: Conf. Ser.* **478** 012011
[5] Basharov A M 2012 *Nanosystems: Physics, Chemistry, Mathematics* **3** 47

Journal of Physics: Conference Series **714** (2016) 012005 doi:10.1088/1742-6596/714/1/012005

# Generation of terahertz radiation in the interaction of coherent electromagnetic wave with a resonant medium in a constant external electric field

**A M Basharov**

Moscow Institute of Physics and Technology, 141700 Dolgoprudny, Russia
National Research Centre «Kurchatov Institute», 123182 Moscow, Russia

E-mail: basharov@gmail.com; http://basharov.me

**Abstract.** The paper shows that it is sufficient to place a resonance medium in an external dc electric field for the generation of low-frequency radiation at the resonant interaction of the medium and a coherent wave with carrier frequency of the optical range. Under certain conditions the carrier frequency of the generated radiation equals the Rabi frequency of the resonant interaction.

## 1. Introduction

Practical applications have encouraged the study of different ways to generate low-frequency radiation, whose carrier frequencies are in the terahertz range. For practical purposes different sources of terahertz radiation are needed - intensive and not really intensive but having the possibility of tuning the carrier frequency. In resonant processes environments with a permanent dipole moment are actively investigated [1-6]. This report shows that low-frequency radiation of low intensity but of easily tunable carrier frequency is not difficult to obtain in any resonance process in environments with no permanent dipole moment when a resonant medium is additionally irradiated with a pulse of a quasipermanent electric field, for example, by placing a resonant medium in the condenser. The generation of low-frequency radiation near the leading edge of the resonant excitation pulse of the optical range is considered and values for the intensity of the generated field and its carrier frequency are obtained. The results follow from consideration of the resonant interaction in view of an external electric field on the basis of the algebraic theory of perturbations.

## 2. Formulation of the task and basic equations

Assume that a coherent electromagnetic wave propagates in the direction of the axis $Z$, with the intensity $E$ of the electric field being represented in the form of a slowly varying amplitude $\mathcal{E}$ and a rapidly changing phase defined by the wave vector $\vec{k} \parallel Z$ and the carrier frequency $\omega$:

$$E = \mathcal{E}\exp[i(kz - \omega t)] + c.c. \equiv \mathcal{E}\exp[-i\Phi] + c.c. \qquad (1)$$

The area of space $0 \le Z \le L$ is filled with the environment consisting of particles of the wave resonant (1) energy levels $|g>$ and $|e>$, with the level $|e>$ being treated as the excited one.

Frequency $\omega_{eg}$ of transition $|e>\to|g>$ is close to frequency $\omega$, $|\omega_{eg}-\omega|<<\omega$, and the very transition is optically permitted, i.e. in states $|g>$ and $|e>$ the dipole matrix element is equal to zero $d_{ee}=d_{gg}=0$.

If one neglects the interaction between particles and relaxation, then the dipole moment of the environment volume unit $P$ can be represented as

$$P=n<\Psi|d|\Psi>, \tag{2}$$

where $n$ is the density of resonant particles in the volume unit, $d=\sum_{kj}d_{kj}|E_k><E_j|$ is the dipole moment operator of a resonant particle, $|E_k>$ is a quantum state of a resonant particle with energy $E_k$, and $|\Psi>$ is its statevector. The polarization of the environment appears in the Maxwell equation, which is to be written in the unidirectional approximation [7]

$$(\frac{\partial}{\partial z}+\frac{1}{c}\frac{\partial}{\partial t})E=-\frac{2\pi}{c}\frac{\partial}{\partial t}P. \tag{3}$$

Neglecting relaxation and spontaneous emission of particles in the electric dipole approximation of the interaction of particles with a coherent field, the Schrödinger equation for the state vector of a resonant particle is conventional

$$i\hbar\frac{d}{dt}|\Psi>=(H_0-Ed-E_cd)|\Psi>,$$

where $H_0$ is the Hamiltonian of an isolated particle, $E_c$ is the intensity of dc field which is additionally applied to the environment.

Standard unitary transformation [8]

$$|\tilde{\Psi}>=e^{-iQ}|\Psi>,\ Q=Q^{+}$$

allows substantiating resonant approximation and defining corrections for the environment polarization with respect to the permanent electric field. The Schrödinger equation for the transformed state vector of resonant atoms: $i\hbar\frac{d}{dt}|\tilde{\Psi}>=\tilde{H}^{(1)}|\tilde{\Psi}>$ is defined by the effective Hamiltonian ($\sigma$ are the Pauli matrices)

$$\tilde{H}^{(1,0)}=\frac{\omega_{eg}}{2}\sigma_3-\mathcal{E}\exp[-i\Phi]d_{ba}\sigma_{+}-\mathcal{E}^{*}\exp[i\Phi]d_{ba}\sigma_{-}$$

and concurs with the equation for the wave function of a two-level particle in the approximation of a rotating wave [8], and environment polarization (2) at low frequencies is defined by the transformed operator of the atomic dipole moment

$$\tilde{d}=e^{-iQ}de^{iQ}=d-i[Q,d]-\frac{1}{2}[Q,[Q,d]]+...=\sum_{kj}\tilde{d}_{kj}|E_k><E_j|. \tag{4}$$

Use is made of the expansion of the transformed Hamiltonian $\tilde{H}$ in powers of constants of interaction of a resonant particle of quantum system with coherent and permanent electric fields whose order is marked by upper indices (the left index is the interaction with a coherent field) [8]:

$$Q(t)=Q^{(1,0)}(t)+Q^{(0,1)}(t)+...,\ \tilde{H}=\tilde{H}^{(1,0)}(t)+\tilde{H}^{(0,1)}(t)+...$$

Low-frequency radiation is defined by the operator ($\omega_{jk} = (E_j - E_k)/\hbar$)

$$Q^{(0,1)} = E_c \sum_{jk}{}' \frac{d_{jk}}{i\hbar\omega_{jk}} | E_j >< E_k |,$$

in consequence of which an atom receives permanent dipolar moments in an electric field

$$\tilde{d}_{ee} = -E_c \sum_j{}' \frac{|d_{ej}|^2}{\hbar\omega_{ej}}, \ \tilde{d}_{gg} = -E_c \sum_j{}' \frac{|d_{gj}|^2}{\hbar\omega_{gj}}. \tag{5}$$

Recall that the states are characterized by a certain parity, that is why in summations (5) $d_{ee} = d_{gg} = 0$ so much as the corresponding terms are equal to zero.

Both permanent electric and resonant coherent fields define the Stark shift of resonant energy levels. Its magnitude is defined by the terms $\tilde{H}^{(2,0)}$ and $\tilde{H}^{(0,2)}$ in the effective Hamiltonian, which, however, should be neglected as they give only corrections for the investigated effect whose emergence is fully defined by the obtained corrections (4), (5). They, in their turn, determine environment polarizations at a low frequency

$$P = n < \tilde{\Psi} | \tilde{d} | \tilde{\Psi} >$$

which provides for low-frequency generation in the process of resonant interaction [5,6].

## 3. Frequency control of the generated low-frequency radiation

It is well known that near the leading edge of the rectangular resonant pulse, as it propagates in the resonant medium, there occur nutational oscillations of pulse intensity due to re-emission of photons in the resonant wavelength. This phenomenon is known as optical nutation and it was repeatedly investigated earlier [8]. At the same time, populations of resonant levels are also subjected to nutational oscillations defined by the Rabi frequency

$$\Lambda = 2 | a d_{eg} / \hbar |,$$

which is the interaction energy of a coherent wave with a resonant particle.

Emergence of the dipole moment of the particles (4), (5) causes oscillations in the polarization of the medium at the Rabi frequency in resonance conditions

$$P_\Lambda = \frac{1}{2} n E_c \left( \sum_j \frac{|d_{ej}|^2}{\hbar\omega_{ej}} - \sum_j \frac{|d_{ej}|^2}{\hbar\omega_{gj}} \right) \cos \Lambda t.$$

In the case of optically thin media of the length $L$, this leads to the following expression for the electric field intensity of the radiation generated at the Rabi frequency beyond the resonant medium

$$E_\Lambda(t,z) = \frac{\pi n \Lambda L}{c} \left( \sum_j \frac{|d_{ej}|^2}{\hbar\omega_{ej}} - \sum_j \frac{|d_{ej}|^2}{\hbar\omega_{gj}} \right) \sin \Lambda \left(t - \frac{z}{c}\right), \tag{6}$$

from which the equality of carrier frequency of low-frequency radiation and the Rabi frequency $\Lambda$ is evident. In order to generate low-frequency radiation in the terahertz range $\Lambda \sim 10^{12}$ $s^{-1}$ exciting resonant fields with the intensity of the order of $I \sim 10^8$ W/cm$^2$ are needed for typical atomic and molecular media. At the same time, an increase in intensity of this radiation can be achieved in ensembles of quantum dots and other artificial objects in which the values of dipole moments are significant, and the frequency of transitions can be reduced during production.

Typically, frequencies of resonance transitions are scattered near some average transition frequency $\omega_0$. Then it is not difficult to generalize (6)

$$E_\Lambda(t,z) = \frac{\pi n L}{c}\left(\sum_j \frac{|d_{ej}|^2}{\hbar\omega_{ej}} - \sum_j \frac{|d_{ej}|^2}{\hbar\omega_{gj}}\right) < \frac{\Lambda}{\Omega}\sin\Omega(t-\frac{z}{c}) > .$$

Here the angle brackets denote averaging over the specified frequency spread with some distribution function $F(\Delta)$, $\Omega = \sqrt{\Delta^2 + \Lambda^2}$.

In general, the carrier frequency of the generated radiation will not be strictly equal to the Rabi frequency and the frequency spectrum occurs near the Rabi frequency. However, importance of this spectrum will be determined by the Rabi frequency quite unambiguously. For example, in the case of symmetrical spread of the Gaussian form, intensity of electric field of low-frequency generation is determined by the Bessel function of order zero:

$$E_\Lambda(t,z) = -\frac{\pi^{3/2} n L \Lambda^2 T_0}{c}\left(\sum_j \frac{|d_{ej}|^2}{\hbar\omega_{ej}} - \sum_j \frac{|d_{ej}|^2}{\hbar\omega_{gj}}\right)\left(J_0(\Lambda(t-\frac{z}{c})) + \Phi_{Error}((t-\frac{z}{c})/T_0) - 1\right).$$

Here $1/T_0$ is the frequency width of spread which is supposed to be much larger than that of the Rabi frequency, $\Phi_{Error}(t)$ is the error function integral.

To experimentally investigate the proposed mechanism of generation of low-frequency radiation, the method of Stark pulse was previously proposed [9,10]. In addition, resonant frequency spread of particles leads to damping of low-frequency radiation despite the fact that the radiation exciting it continues to exert on the resonance environment. Here the signals of low-frequency radiation will emerge near the front and near the edges of the Stark pulses of a permanent electric field. Emergence of a nutational echo signal [10] at low frequency is also possible, which can simultaneously serve as a source of a variety of spectroscopic information. Another way of providing effective resonant interaction of medium with an exciting field of allegedly rectangular shape is the method of tuning the exciting radiation frequency [11,12].

**Acknowledgments**
The author expresses his gratitude to S.V.Sazonov for many helpful discussions. This work is supported in part by the Russian Foundation for Basic Research (grant No. 16-02-00453a).

**References**

[1] Lee Yun-Shik 2009 *Principles of terahertz science and technology* (Springer Science + Business Media)
[2] Kmetic M A and Meath W J 1985 *Phys. Lett.* A **108** 340
[3] Lachaine J M *et al* 2000 *Phys. Rev.* B **62** R4829
[4] Bugai A N and Sazonov S V 2014 *JETP Lett.* **98** 638
[5] Kibis O V *et al* 2009 *Phys. Rev. Lett.* **102** 023601
[6] Basharov A M 2016 *JETP Lett.* **103** 16
[7] Belenov E M and Nazarkin A V 1990 *JETP Lett.* **51** 288
[8] Maimistov A I and Basharov A M 1999 *Nonlinear optical waves* (Dordrecht: Kluwer Academic)
[9] Brewer R G and Shoemaker R L 1971 *Phys. Rev. Lett.* **27** 631
[10] Alekseev A I and Basharov A M 1979 *Sov. Phys. JETP* **50** 272
[11] Alekseev A I and Basharov A M 1980 *Phys. Lett.* A **77** 123
[12] Alekseev A I and Basharov A M 1982 *J. Phys.* B **15** 4269

Journal of Physics: Conference Series **714** (2016) 012006 doi:10.1088/1742-6596/714/1/012006

# Investigation of strain-induced magnetization change in ferromagnetic microparticles

**A P Chuklanov[1], N I Nurgazizov[1,2], D A Bizyaev[1], T F Khanipov[1], A A Bukharaev[1,2], V Yu Petukhov[1,2], V V Chirkov[1] and G G Gumarov[1]**

[1] Kazan E. K. Zavoisky Physical -Technical Institute, 420034, 10/7 Sibirskiy Trakt, Kazan, Russian Federation

[2] Kazan Federal University, 420008, 8 Kermlevsakya St, Kazan, Russian Federation

E-mail: achuklanov@kfti.knc.ru

**Abstract.** This work is devoted to investigation of magnetoelastic strain effect on the ferromagnetic microparticles of permalloy. An original method of sample fabrication with compressed microparticles is proposed. Magnetic force microscopy and magneto-optical Kerr experiments were carried out with unstrained and compressed microparticles. The domain walls transformation in compressed microparticles is in good agreement with numerical calculations. Hard axis of magnetization was observed on the compressed sample.

## 1. Introduction

Studying of the remagnetization processes caused by the magnetoelastic effect has a big interest because of ability to manage the magnetization by an electric field [1]. It is caused by the demand of increasing data density of magnetic media, speeding up the read-write operations and reliability of the storage. The transition to recording information by electric field is mainly interesting because a very low energy is required for recording a bit of information (about 0.5fJ). This is several orders lower than for other known methods of information recording [2]. This method is called MERAM – magnetoelectric random access memory. For example, if one can deposit a magnetic layer (a continuous film or particles) on the surface of a piezoelectric material, then applying an electric field may cause mechanical deformation and change the magnetic state of the upper layer [3]. Inducing mechanical stresses in the magnetic structures is also possible by depositing it on pre-curved elastic substrates. After substrate straightening mechanical stress occurs and uniaxial anisotropy is induced [4].

In this work the changes of the magnetization structure and coercive properties of permalloy microparticles under elastic mechanical stresses are studied by magnetic force microscopy (MFM) and magneto-optical Kerr-effect (MOKE).

## 2. Results and Discussion

In this paper a magnetoelastic phenomenon of an aligned ensemble of permalloy ($Ni_{75}Fe_{25}$) microparticles with size 22×22×0.03 $mkm^3$ detached from each other on 10 mkm is studied. The elastic tension is created by a method described in [4]. Two types of samples were fabricated (figure 1). For the samples of the first type, a metal grid with square holes of 22×22 $mkm^2$ was placed on the flat (not curved) glass substrate (figure 1c). For the samples of the second type the metal grid was placed on the curved surface of the substrate. The bending was carried out by placing the metal wire with diameter of 80 mkm under the substrate and fixing the edges of the substrate by holders. The permalloy (Py) layer with 30 nm thickness was simultaneously deposited on both samples in ultrahigh vacuum by "Multiprobe P" device (Omicron). After ending of the deposition the samples were extracted from the holders in the atmosphere conditions. It is obvious that the Py microparticles experienced a uniaxial compressive strain after straightening the substrate in the samples of the second type. The image of the part of the structure obtained by an atomic force microscope Solver P47 Pro (NT-MDT) is shown in figure 1d.

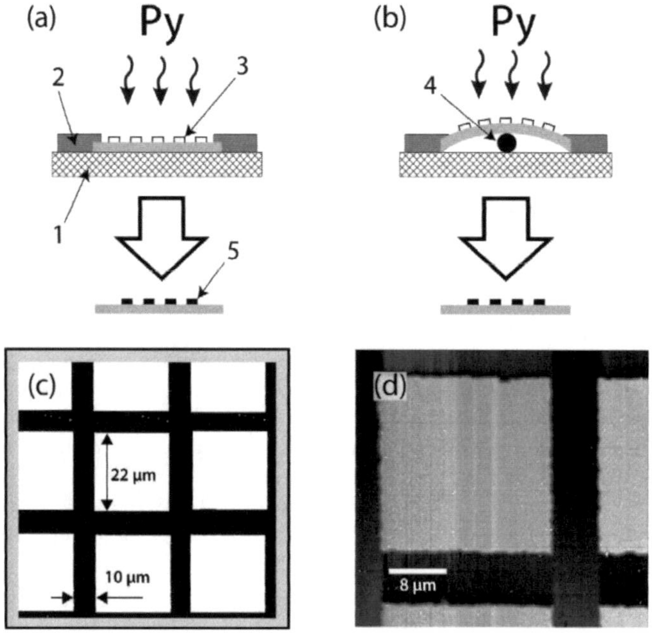


**Figure 1.** a – scheme of fixing lithographic mask (metal grid) on the original (flat) substrate; b – scheme of fixing lithographic masks on the curved substrate; c – scheme of lithographic mask; d – AFM-image of planar structure of the Py microparticles. 1 – base; 2 – holders; 3 – metal grid; 4 – microwire; 5 – Py microparticles.

The changes of the magnetization structure induced by mechanical stresses have been studied by Solver P47 Pro in MFM mode. The MFM-image of Py microparticles at the initial state has a typical structure for permalloy particles of this size and shape and shows the presence of four domains of the same shape, separated by cross-like domain walls (figure 2a). Under the compression, the domain walls in a particle are deformed and a characteristic bridge at the center of the particle appears (figure 2b). In order to compare the MFM-images with the magnetization distribution in the particle a computer modeling program OOMMF [5] was used. It is based on the numerical solution of the Landau-Lifshitz-Gilbert equation, corresponding to the distribution of local magnetic moments with a minimum total energy. In the simulation, the size of the particle was fixed and the saturation magnetization ($M_0$), and magnetoelastic anisotropy ($H_k$) was varied. After that, using the calculated magnetization distribution the MFM-image was simulated (by computer program "Virtual MFM" [6] developed in our laboratory) and compared with the experimental MFM-image (figure 2a,b).

Simulated MFM-images closest to experiment were obtained at $H_k$ = 7 Oe, $M_0$ = 889 Oe for the sample of the first type (uncompressed) and $H_k$ = 42 Oe, $M_0$ = 895 Oe for the sample of second type (compressed). The same parameters were experimentally obtained for the same samples from an analysis of the angular dependence of the spectra of ferromagnetic resonance (FMR) [7]. In our opinion, a nonzero value of $H_k$ for the sample of the first type is caused of minor residual stresses arising during the deposition. Figure 2 shows good agreement between the experimental and modeling MFM-images.

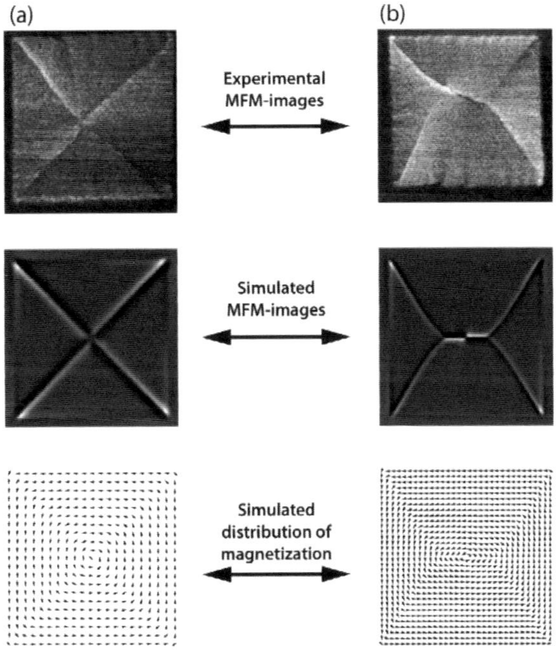


**Figure 2.** The experimental and simulated MFM-images and distribution of magnetization of Py microparticle in the unstrained (a) and compressed (b) states.

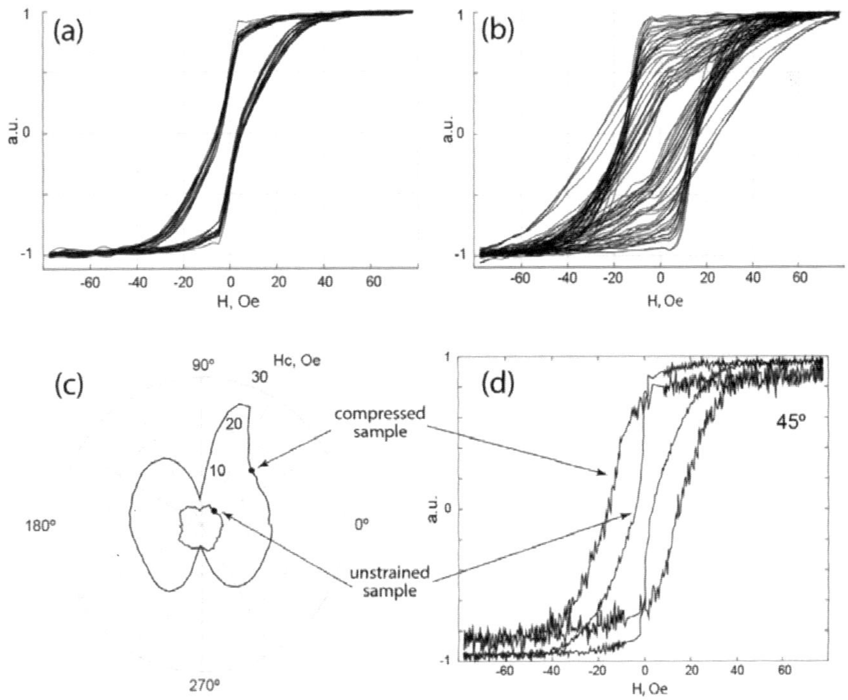


**Figure 3.** MOKE experiments. The hysteresis loops of Py microparticles obtained at different azimuthal angles for the relaxed (a) and compressed (b) samples. c – the angular dependence for coercivity force of unstrained and

compressed microparticles; d – the selected hysteresis loops for angle 45 deg.

The influence of uniaxial compression on the magnetic structure of microparticles was investigated by scanning polarimetry device, based on the ellipsometer. The device is based on magneto-optical Kerr effect (MOKE) and allows to register hysteresis loops (from -100 Oe to +100 Oe) at different azimuthal angles between magnetic field and sample. In figure 3 the hysteresis loops of the samples of the first and second types under different azimuthal angles (i.e rotation takes place in the plane of the substrate) and the angular dependences of coercivity force ($H_c$) are shown. From these dependencies, one can see that due to the magnetoelastic effect the sample with microparticles under compression has a preferred direction (figure 3c) and $H_c$ varies from 7 to 25 Oe. The sample has a hard axis of magnetization because magnetization saturation exceeds 70 Oe along the selected direction (figure 3b).

## 3. Conclusion

By the analysis of experimental data, one can see that significant changes the magnetic properties of Py microparticles under mechanical stress appeared due to magnetoelastic effect. The changes of four-domain magnetic structure registered by MFM and MOKE data demonstrate the strong angular dependence of coercivity force and magnetization saturation.

## Acknowledgments

The authors would like to thank Ju. A. Kandrashkin for the calculations of magnetic anisotropy and magnetization saturation based on FMR spectra. This work is supported by "Russian Foundation for Fundamental Research" (RFFI No. 15-02-02728) and the program of Presidium Russian Academy of Sciences.

## References

[1] Morozov A I 2014 *Physics of the Solid State* **56** 865–872

[2] Hockel J L Deterministic Magnetization Control by Magnetoelastic Anisotropy and its Dependence on Geometry and Scale 2013 Ph.D. Theses (UCLA Electronic Theses and Dissertations).

[3] Chung T K, Wong K, Keller S, Wang K L and Carman G P 2009 *J. Appl. Phys.* **106** 103914

[4] Belyaev B A and Izotov A V 2007 *Physics of the Solid State* **49** 1731–1739

[5] Donahue M J and Porter D G Object oriented micromagnetic framework (OOMMF) (http://math.nist.gov/oommf/)

[6] Ovchinnikov D V and Bukharaev A A 2001 *Technical Physics* **46** 1014–19

[7] Biziyaev D A, Bukharaev A A, Kandrashkin Yu E, Gorev R V, Mironov V L, Nurgazizov N I and Khanipov T A 2015 *Proc. Int. Conf. Modern Development of Magnetic Resonance* (*Kazan, Russia, 22-26 September 2015*) pp 54–55

Journal of Physics: Conference Series **714** (2016) 012007 doi:10.1088/1742-6596/714/1/012007

# Retardation effect in the recoil corrections to the fine shift of $S$ energy levels of hydrogen-like atoms

**S Churochkina and A Udalova**

Saratov State University, 410012, Astrakhanskaya St. 83, Saratov, Russia

E-mail: klechshevskaya@mail.ru

**Abstract.** Contributions of recoil effects to the fine shift of $S$ energy levels of hydrogen-like atoms have been considered. The detailed analysis of the influence of retardation effect on the value of the logarithmic in respect to the mass ratio correction from the two photon diagram of sequential exchange by Coulomb and transverse photons has been realized.

## 1. Introduction

The problem of refinement of calculations of recoil corrections for the energy levels of hydrogen-like atoms (HLA) has attracted a lot of attention during last few years [1–9]. Being simplest possible bound state in quantum electrodynamics (QED) the HLA allows to verify experimentally all fundamental QED predictions. On the other hand, recently, owing to the breakthrough in the Doppler-free two-photon spectroscopy [10] the value of the frequency of key for the theory of $1S$-$2S$ transition in the hydrogen atom has been evaluated with the precision up to 10th Hz [11]

$$f_{1S-2S} = 2\,466\,061\,413\,187\,018\,(11)\ \text{Hz}. \tag{1}$$

And, it certainly challenges all theoretical QED predictions [1,8,9]. That is why, a few extended reviews (see for example [9]) devoted to the theoretical description of the HLA have been published. In this regards, the problem of the achievement of the consistency between the modern experimental measurements of the fine shift of energy levels of HLA and the corresponding theoretical results has been marked as one out of the most stimulating issues for the theory.

Indeed, it has been established that neither the Dirac equation nor the Schrödinger one suffice for the full description of the HLA, because of its bound state nature. In this case, the QED field-theoretical methods [12–17] developed for bound states required to describe such phenomena as: Lamb shift, both fine shift and hyperfine splitting corrections due to the radiation and recoil effects. It is worth mentioning that, historically, the treatment of the two particle bound state problem had been done firstly on the ground of the Bethe-Salpeter equation [12]. Afterwards, different methods based on the field-theoretical quasipotential approach (QA) emerged [18, 19].

As it is known, all electrodynamic corrections to the energy levels of HLA may be written in the form of the series expansion in powers of three small parameters $\alpha$, $Z\alpha$ and $m_1/m_2$ (where $\alpha$ is the fine structure constant, $Z$ is the charge of nucleus being considered, $m_1$ and $m_2$ are the masses of the light and the heavy particle, respectively) which determine the properties of the bound state. Corrections which depend on $Z\alpha$ and $m_1/m_2$ are called recoil corrections.

In particular, in this work we focus on the calculation of recoil corrections to the fine shift of $S$ energy levels of HLA on the ground of QA. For the first time this problem was treated in [20]. The result was,

$$\Delta E = \frac{1}{\pi}\frac{(Z\alpha)^5\mu^3}{m_1 m_2}\frac{1}{n^3}\left[A + \frac{2}{3}\delta_{l0}\ln{(Z\alpha)}^{-1} - \frac{8}{3}\ln[k_0(n)] + \right. \\ \left. + \frac{2}{{m_2}^2 - {m_1}^2}\delta_{l0}\left({m_1}^2\ln\frac{m_2}{\mu} - {m_2}^2\ln\frac{m_1}{\mu}\right)\right], \tag{2}$$

where $n$ is the principal quantum number, $l$ is the orbital quantum number, $A$ is constant, $\ln[k_0(n)]$ is the Bethe logarithm, $\mu = \frac{m_1 m_2}{m_1+m_2}$ is the reduced mass.

This recoil correction $\Delta E$ is generated by the kern of interactions between two photons. In expression (2) there is a contribution linear in the logarithm of the mass ratio of particles

$$\Delta E^{\ln\beta} = \frac{2}{\pi}\frac{(Z\alpha)^5\mu^3}{m_1 m_2}\frac{\beta^2}{n^3(1-\beta^2)}\ln\beta^{-1}, \qquad \beta = \frac{m_1}{m_2}. \tag{3}$$

It should be emphasized that, the evaluation of other logarithmic on the mass ratio contributions to the fine shift of energy levels of the HLA has not been counted entirely yet [21–29]. In our opinion, the improvement achieved in the precision of the measurement of the $1S$-$2S$ transition frequency increases the value of such estimation. And, even going deeply, in this article, we calculate the influence of the retardation effect on the value of the logarithmic in respect to the mass ratio correction from the two photon diagram of sequential exchange by Coulomb and transverse photons.

## 2. Retardation effect

The relativistic modification of the Coulomb interaction has to be required under the investigation of recoil effects in the HLA up to the order $\frac{(Z\alpha)^6\mu^3}{m_1 m_2}\beta\ln\beta^{-1}$, i.e. integrals of the following type should be considered

$$i = \int\frac{d^3q}{(q^2+\alpha^2\mu^2)^2\mathbf{k_q}(\mathbf{k_q}+s)}, \qquad \mathbf{k_q} = |\mathbf{k}-\mathbf{q}|. \tag{4}$$

The integral from (4) with $s=0$ gives the integral with Coulomb interaction

$$I = \int\frac{d^3q}{(q^2+\alpha^2\mu^2)^2{\mathbf{k_q}}^2}. \tag{5}$$

After integrating with respect to angles in (4) and (5) they take the form:

$$i = \frac{2\pi}{k}\int\limits_0^\infty\frac{qdq}{(q^2+\alpha^2\mu^2)^2}\ln\frac{k+q+s}{|k-q|+s}, \qquad I = \frac{2\pi}{k}\int\limits_0^\infty\frac{qdq}{(q^2+\alpha^2\mu^2)^2}\ln\frac{k+q}{|k-q|}. \tag{6}$$

Using the well-established properties of integrals one can conclude that for any $k$ and $s$ the condition $i < I$ should be fulfilled.

In turn, corrections to the Coulomb energy levels can be found by calculating integrals containing expressions from (6) as a part of its integrands along with the parameter $k$ as the variable of integration.

For a start, let's analyze the result of the integration of expressions from (6) with respect to momentum $q$

$$i=\frac{2\pi}{k}\left[\frac{1}{4}\ln\frac{k^2+\alpha^2\mu^2}{s^2}\left(\frac{1}{(k+s)^2+\alpha^2\mu^2}-\frac{1}{(k-s)^2+\alpha^2\mu^2}\right)+\right.$$
$$+\frac{\pi}{4\alpha\mu}\left(\frac{k+s}{(k+s)^2+\alpha^2\mu^2}+\frac{k-s}{(k-s)^2+\alpha^2\mu^2}\right)+$$
$$\left.+\left(\frac{k+s}{(k+s)^2+\alpha^2\mu^2}-\frac{k-s}{(k-s)^2+\alpha^2\mu^2}\right)\frac{1}{2\alpha\mu}\arctan\frac{k}{\alpha\mu}\right], \tag{7}$$

$$I=\frac{\pi^2}{\alpha\mu}\frac{1}{k^2+\alpha^2\mu^2}. \tag{8}$$

Now, in accordance with general rules of estimation of integrals, we initially consider everywhere the leading contribution with $I$ $(s=0)$ in integrands. If it leads to the contribution proportional to $\frac{(Z\alpha)^6\mu^3}{m_1m_2}\beta\ln\beta^{-1}$ then the corresponding integral with $i$ $(s\neq 0)$ should be counted as well.

As an example, we treat the Feynman diagram with parallel photon lines corresponding to the exchange by one Coulomb and one transverse photons.

The contribution to the fine shift from sequential exchange by Coulomb and transversal photons is defined by the expression

$$\Delta E_{\rm par}=\langle\Psi'_{nS}|(K_{\rm C}G_0K_T+K_TG_0K_{\rm C})^+_{0F}|\Psi'_{nS}\rangle, \tag{9}$$

where $G_0$ is a Green function of free fermions, $K_{\rm C}$ and $K_T$ are kernels, describing the exchange by one Coulomb and one transversal photons respectively, symbol $(...)^+_{0F}=F^{-1}\overline{G_0(...)G_0}^+F^{-1}$, $F=(2\pi)^3\delta^3(\mathbf{p}-\mathbf{q})(E-\varepsilon_{1p}-\varepsilon_{2p})^{-1}$, $\varepsilon_{ip}=\sqrt{p^2+m_i^2}$, $i=1,2$, $E$ is the total energy of a system of two relativistic particles, $\Psi'_{nS}(\mathbf{p})=\Omega_p\Psi_{nS}(\mathbf{p})$, $\Omega_p$ is a factor, emerged because of the relativistic character of the quasipotential equation, $\Psi_{nS}$ is a wave function which defines the state of a two-fermion system with a total energy $E$.

In particular, the logarithmic contribution to the fine shift of the $1S$ energy level of the HLA (with $Z=1$) appeared from the following integral

$$\Delta E_{\rm par}(\ln^2\beta^{-1})=-\frac{\alpha^7\mu^5}{2\pi^6}\int\frac{d^3pN_p\Omega_p}{(p^2+\alpha^2\mu^2)^2}\int\frac{d^3qN_q\Omega_q}{(q^2+\alpha^2\mu^2)^2}\times$$
$$\times\int\frac{d^3kk^2}{\varepsilon_{1k}\varepsilon_{2k}(\mathbf{p}-\mathbf{k})^2\mathbf{k_q}(\mathbf{k_q}+\varepsilon_{1k}+\varepsilon_{1q})(\varepsilon_{1k}+\varepsilon_{2k}+E_1+E_2)}, \tag{10}$$

where

$$N_p=\sqrt{\frac{(\varepsilon_{1p}+m_1)(\varepsilon_{2p}+m_2)}{2\varepsilon_{1p}\varepsilon_{2p}}},\qquad \Omega_p=\frac{(\varepsilon_{1p}+m_1)(\varepsilon_{2p}+m_2)}{2\mu(\varepsilon_{1p}+\varepsilon_{2p}+E_1++E_2)}.$$

Leaving in (10) only terms proportional to $\frac{\alpha^6\mu^3}{m_1m_2}\beta\ln^2\beta^{-1}$, we got

$$I_1=-\frac{\alpha^7\mu^3}{8\pi^6}m_1\left\{\int\frac{d^3p}{(p^2+\alpha^2\mu^2)(\varepsilon_{1p}+m_1)}\int\frac{d^3q}{(q^2+\alpha^2\mu^2)^2}\int\frac{d^3kk^2}{\mathbf{k_p^2k_q}}\times\right.$$
$$\times\frac{1}{\varepsilon_{1k}\varepsilon_{2k}(\mathbf{k_q}+\varepsilon_{1k}+E_1)(\varepsilon_{1k}+\varepsilon_{2k}+E_1+E_2)}+\int\frac{d^3p}{(p^2+\alpha^2\mu^2)^2}\int\frac{d^3q}{(q^2+\alpha^2\mu^2)(\varepsilon_{1q}+m_1)}\times$$
$$\left.\times\int\frac{d^3kk^2}{\varepsilon_{1k}\varepsilon_{2k}\mathbf{k_p^2k_q}(\mathbf{k_q}+\varepsilon_{1k}+E_1)(\varepsilon_{1k}+\varepsilon_{2k}+E_1+E_2)}\right\}=I_1'+I_1''. \tag{11}$$

Let's focus on the first term from (11). By counting integrations with respect to momenta $\mathbf{p}$ and $\mathbf{q}$ it can be rewritten as

$$I_1' = -\frac{\alpha^7\mu^3}{2\pi^5}m_1\int\limits_0^\infty \frac{dk k^4}{\varepsilon_{1k}\varepsilon_{2k}(\varepsilon_{1k}+\varepsilon_{2k}+E_1+E_2)}j_p j_q, \tag{12}$$

where

$$j_p = \int\frac{d^3p}{(p^2+\alpha^2\mu^2)(\varepsilon_{1p}+m_1)\mathbf{k_p}^2} = \frac{2\pi}{k^2+\alpha^2\mu^2}\left(-1+\frac{\varepsilon_{1k}}{k}\ln\frac{\varepsilon_{1k}+k}{m_1}\right)+\frac{\pi^2}{m_1 k}\arctan\frac{k}{\alpha\mu}+$$
$$+\frac{2\pi}{k}\left(\frac{k}{3\varepsilon_{1k}^2}+\frac{1}{\varepsilon_{1k}}\left(1+\frac{m_1^2}{3\varepsilon_{1k}^2}\right)\ln\frac{\varepsilon_{1k}+k}{m_1}\right),$$

$$j_q = \int\frac{d^3q}{(q^2+\alpha^2\mu^2)^2\mathbf{k_q}(\mathbf{k_q}+\varepsilon_{1k}+E_1)} = \frac{\pi}{km_1}\left(\frac{1}{\alpha\mu}\arctan\frac{k}{\alpha\mu}-\frac{\pi}{2\alpha\mu}\frac{k}{\varepsilon_{1k}+m_1}+\right.$$
$$\left.+\frac{k}{2m_1(\varepsilon_{1k}+m_1)}\ln\frac{\varepsilon_{1k}+E_1}{\varepsilon_{1k}-E_1}\right).$$

It is worth mentioning here, that $j_p$ from (12) contains the Coulomb factor $\frac{1}{\mathbf{k_p}^2}$ whereas $j_q$ holds its relativistic modification $\frac{1}{\mathbf{k_q}(\mathbf{k_q}+\varepsilon_{1k}+E_1)}$. And, if we make the replacement $\mathbf{k_q}+\varepsilon_{1k}+E_1 \to \mathbf{k_q}$ in $j_q$, then, one can easily verify that, $j_q$ reduces to $I$ from (8). It leads to the condition $j_q < I$. Following the above-mentioned criterium we must evaluate the integral $I_1'$ because its leading contribution with $I$ instead of $j_q$ is nonzero.

Evaluating the integral $I_1'$ up to the order of $\frac{\alpha^6\mu^3}{m_1 m_2}\beta\ln^2\beta^{-1}$ we come to the integral

$$I_1' = -\frac{1}{\pi^2}\frac{\alpha^6\mu^3}{m_1m_2}m_1(m_1+m_2)\int\limits_0^\infty\frac{dk}{\varepsilon_{2k}(\varepsilon_{1k}+m_1)(\varepsilon_{2k}+m_2)}\ln\frac{\varepsilon_{1k}+k}{m_1}. \tag{13}$$

After calculating it we have

$$I_1'(j_q) = -\frac{1}{4\pi^2}\frac{\alpha^6\mu^3}{m_1m_2}\beta\ln^2\beta^{-1}. \tag{14}$$

If we use in (12) $I$ instead of $j_q$ then, after passing through the same steps we obtain

$$I_1'(I) = -\frac{2}{4\pi^2}\frac{\alpha^6\mu^3}{m_1m_2}\beta\ln^2\beta^{-1}. \tag{15}$$

Therefore, the relativistic modification of the Coulomb interaction reduces its effect twice in comparison with the non-relativistic case.

The second term in (11) can be evaluated following the same procedure as it was used for the estimation of the first one. It yields, in the relativistic case

$$I_1'' = -\frac{1}{2\pi^2}\frac{\alpha^6\mu^3}{m_1m_2}\beta\ln^2\beta^{-1}. \tag{16}$$

Thus, the total correction to the fine shift of $1S$ energy level of the HLA from the contribution presented by (10) takes the form

$$\Delta E_{\mathrm{par}}(\ln^2\beta^{-1}) = I_1 = -\frac{3}{4\pi^2}\frac{\alpha^6\mu^3}{m_1m_2}\beta\ln^2\beta^{-1}. \tag{17}$$

For comparison, the non-relativistic version of the same correction can be written as

$$\Delta E_{\mathrm{par}}(J_q) = I_1 = -\frac{1}{\pi^2}\frac{\alpha^6\mu^3}{m_1m_2}\beta\ln^2\beta^{-1}. \tag{18}$$

## 3. Conclusion

Therefore, we can conclude, that the relativistic modification of the Coulomb interaction reduces the value of $\Delta E$. It should be counted in further calculations of corrections to energy levels of the HLA.

## References

[1] Mohr P J, Taylor B N and Newell D B 2012 *Rev. Mod. Phys.* **84** 1527
[2] Karshenboim S G, Pavone F S, Bassani F, Inguscio M and Hänsch T W 2001 *The Hydrogen Atom: Precision Physics of Simple Atomic Systems* vol 570 (Berlin, Heidelberg, New York, Barcelona, Hong Kong, London, Milan, Paris, Singapore, Tokyo: Springer)
[3] Karshenboim S G 2008 *Precision Physics of Simple Atoms and Molecules* vol 745 (Berlin, Heidelberg: Springer)
[4] Dvoeglasov V V, Tyukhtyaev Yu N and Faustov R N 1994 *Fiz. Elem. Chastits At. Yadra* **25** 144 [*Phys. Part. Nucl.* **25** 58]
[5] Kinoshita T 1996 *Rep. Prog. Phys.* **59** 1459
[6] Sapirstein J 2006 *Handbook of Atomic, Molecular and Optical Physics*, ed Drake G W F (Heidelberg: Springer) p 413
[7] Mohr P J and Taylor B N 2006 *Handbook of Atomic, Molecular and Optical Physics*, ed Drake G W F (Heidelberg: Springer) p 429
[8] Eides M I, Grotch H and Shelyuto V A 2001 *Phys. Rep.* **342** 63
[9] Eides M I, Grotch H and Shelyuto V A 2007 *Theory of light hydrogenic bound states* Springer Tracts Mod. Phys. **222** (Berlin, Heidelberg, New York: Springer)
[10] Hänsch T W 2006 *Rev. Mod. Phys.* **78** 1297
[11] Matveev A, Parthey Ch G, Predehl K, Alnis J, Beyer A, Holzwarth R, Udem Th, Wilken T, Kolachevsky N, Abgrall M *et al* 2013 *Phys. Rev. Lett.* **110** 230801
[12] Bethe H A and Salpeter E E 1951 *Phys. Rev.* **84** 1232
[13] Logunov A A and Tavkhelidse A N 1963 *Nuovo Cimento* **29** 380
[14] Kadyshevsky V G 1968 *Nucl. Phys.* B **6** 125
[15] Grotch H and Yennie D R 1969 *Rev. Mod. Phys.* **41** 350
[16] Gross F 1969 *Phys. Rev.* **186** 1448
[17] Dulyan L S and Faustov R N 1975 *Teor. Mat. Fiz.* **22** 314 [*Theor. Math. Phys.* **22** 220]
[18] Faustov R N 1964 *JINR preprint* P-1572 (Dubna: JINR)
[19] Tyukhtyaev Yu N 1982 *Teor. Mat. Fiz.* **53** 419 [*Theor. Math. Phys.* **53** 1217]
[20] Fulton T and Martin P C 1954 *Phys. Rev.* **95** 811
[21] Boikova N A, Tyukhtyaev Yu N and Faustov R N 1998 *Yad. Fiz.* **61** 866 [*Phys. At. Nucl.* **61** 781]
[22] Boikova N A, Tyukhtyaev Yu N and Faustov R N 2001 *Yad. Fiz.* **64** 986 [*Phys. At. Nucl.* **64** 917]
[23] Boikova N A, Kleshchevskaya S V, Tyukhtyaev Yu N and Faustov R N 2003 *Yad. Fiz.* **66** 925 [*Phys. At. Nucl.* **66** 893]
[24] Boikova N A, Nunko N E, Tyukhtyaev Yu N and Faustov R N 2002 *Teor. Mat. Fiz.* **132** 339 [*Theor. Math. Phys.* **132** 1179]
[25] Boikova N A, Kleshchevskaya S V, Tyukhtyaev Yu N and Faustov R N 2006 *Teor. Mat. Fiz.* **149** 325 [*Theor. Math. Phys.* **149** 1577]
[26] Boikova N A, Boikova O A, Kleshchevskaya S V and Tyukhtyaev Yu N 2008 *Izv. Saratov. un-ta. Novaya seriya. Fizika* **8** 42
[27] Boikova N A, Kleshchevskaya S V, Tyukhtyaev Yu N and Faustov R N 2009 *Yad. Fiz.* **72** 300 [*Phys. At. Nucl.* **72** 272]
[28] Boikova O A, Kleshchevskaya S V, Tyukhtyaev Yu N and Faustov R N 2010 *Yad. Fiz.* **73** 1024 [*Phys. At. Nucl.* **73** 988]
[29] Churochkina S V 2014 *Geteromagnitnaya mikroelektronika* **17** 23

Journal of Physics: Conference Series **714** (2016) 012008 doi:10.1088/1742-6596/714/1/012008

# The influence of the nuclear self energy function on the spectral lines of atoms with unstable nuclei

**R Kh Gainutdinov, A A Mutygullina and A S Petrova**
Kazan Federal University, 18 Kremlyovskaja St., 420008, Kazan, Russia
E-mail: Renat.Gainutdinov@kpfu.ru

**Abstract.** Optical emission from atoms with radioactive nuclei is investigated. The self energy function of the unstable nucleus is constructed. A new arbitrary parameter characterizing the nuclear interaction is introduced. The instability of the atomic nucleus is shown to result in the essential broadening of optical emission lines. The spectral line shape is shown to be of the Lorentz form, with the width being a nonlinear combination of the nuclear and atomic decay widths multiplied by the parameter depending on the peculiarities of nuclear interaction.

## 1. Introduction

Extremely high intensity and short duration of pulses achieved by modern laser spectroscopy have enabled registering spectra of atoms with exotic and short-living nuclei. Recent systematic investigation of fluorescence spectra of isotopic chains of heavy elements has lead to the discovery of regularities in the nuclear mean square charge radius [1] and nuclear magnetic anomaly [2]. To interpret these phenomena an improvement of the existing nuclear models and the theory of effective forces in nuclear physics is required. Investigating atoms of short-living isotopes we should pay attention to the fact that nuclear radioactivity itself has an effect on the look of atomic spectrum. Usually, nuclear radioactivity is taken into account by simple addition of the nuclear decay width and the decay width of an atomic energy level. However, this is a rough approximation. In this paper a consistent derivation of the equation for an emission line of an atom with a radioactive nucleus is presented. For this aim the Green's function formalism is used, which is a widespread technique in theoretical quantum physics [3]. The self energy function of the atomic nucleus is constructed in the framework of the generalized quantum dynamics formalism [4, 5] which has been successfully applied for many problems of atomic and nuclear physics (see for example [6, 7, 8]). A new parameter characterizing the dynamics of nuclear decay is introduced. The emission line shapes are shown to be of the Breit-Wigner form, with the widths depending on the above parameter and equaling the simple sum of nuclear and electronic decay widths in the case when the parameter equals zero.

## 2. Decay of an atom with an unstable nucleus

Let us consider an atom consisting of an unstable nucleus of charge $Z$ and $Z$ electrons. The atom transits from the excited state $|i\rangle$ with energy $E_i$ into the ground state $|f\rangle$ with energy $E_f$ emitting a photon with momentum $\mathbf{k}$, polarization $\varepsilon_\lambda$ and energy $\omega$, afterwards its nucleus decays. In this paper we restrict ourselves to the case of a two-body nuclear decay. Hence, the final state $|m_1, m_2; \mathbf{k}, \varepsilon_\lambda\rangle$ will contain two new atoms with masses $m_1$, $m_2$, momenta $\mathbf{p}$, $-\mathbf{p}$ (in the center of mass frame) and the numbers of electrons $Z_1$, $Z_2$ ($Z_1 + Z_2 = Z$), as well as the photon emitted on the optical transition $|i\rangle \to |f\rangle$. The spectral contour of the line, emitted in this process, is determined by the probability $dW$, that the energy of the photon lies in the interval $(\omega, \omega + d\omega)$. This probability is the absolute square of the corresponding matrix element of the evolution operator $U(t, 0)$ in the limit of infinite times $t$ [9]:

$$\mathrm{d}W = \lim_{t \to 0} \int \frac{\mathrm{d}\mathbf{p}}{(2\pi)^{3/2}} |\langle m_1, m_2; \mathbf{k}, \varepsilon_\lambda | U(t,0) | i \rangle|^2 \, \mathrm{d}\mathbf{k}, \tag{1}$$

In [5] it was shown that the evolution operator can be defined as Fourier transform of the full Green operator $G(z)$:

$$U(t,0) = \frac{1}{2\pi} \int G(z) e^{-izt} \mathrm{d}z. \tag{2}$$

For the system under study the full Green operator looks as follows:

$$G(z) = G_0^{(dp)}(z) M^{(\mathrm{N})}(z) G_0^{(f)}(z) M^{(\mathrm{em})}(z) G_0^{(i)}(z), \tag{3}$$

where

$$G_0^{(i)}(z) = \frac{|i\rangle\langle i|}{z - E_i - C(z)} \tag{4}$$

is the Green operator of the initial atom in the excited state,

$$G_0^{(f)}(z) = \sum_\lambda \int \frac{|f; \mathbf{k}, \varepsilon_\lambda\rangle\langle f; \mathbf{k}, \varepsilon_\lambda|}{z - E_f - \omega - C(z - \omega)} \mathrm{d}\mathbf{k}, \tag{5}$$

is the Green operator of the intermediate state $|f; \mathbf{k}, \varepsilon_\lambda\rangle$ containing the de-excited atom and the photon, and

$$G_0^{(dp)}(z) = \frac{|m_1, m_2; \mathbf{k}, \varepsilon_\lambda\rangle\langle m_1, m_2; \mathbf{k}, \varepsilon_\lambda|}{z - E_{dp} - \omega} \tag{6}$$

is the Green operator of the final system, consisting of the decay products with the total energy $E_{dp}$ and the photon. $M^{(\mathrm{em})}(z)$ and $M^{(\mathrm{N})}(z)$ are operators of electromagnetic and nuclear interaction respectively, and $C(z)$ is an operator of the interaction with vacuum.

Substituting (4), (5) and (6) into (2) and taking the limit $t \to \infty$, we obtain for the matrix element of the evolution operator:

$$\begin{aligned} S_{fi} &= \lim_{t \to 0} \langle m_1, m_2; \mathbf{k}, \varepsilon_\lambda | U(t,0) | i \rangle \\ &= A\omega \frac{M_{N\mu}}{E_{dp} - E_f - C_f(E_{dp})} \frac{M_{fi}}{E_{dp} - E_i + \omega - C_i(E_{dp} + \omega)}, \end{aligned} \tag{7}$$

where $M_{fi} = \langle \varphi_f; \mathbf{k}, \varepsilon_\lambda | M^{(\mathrm{em})}(E_f + \omega) | \varphi_i \rangle$ and $M_{N\mu} = \langle m_1, m_2; \mathbf{k}, \varepsilon_\lambda | M^{(\mathrm{N})}(E_f + \omega) | \varphi_f; \mathbf{k}, \varepsilon_\lambda \rangle$ are the matrix elements of the electromagnetic and nuclear interaction operators respectively; $C_i(z) = \langle i | C(z) | i \rangle$ and $C_f(z) = \langle f; \mathbf{k}, \varepsilon_\lambda | C(z) | f; \mathbf{k}, \varepsilon_\lambda \rangle$ are the self energy functions of the atom in the excited and de-excited states correspondingly; and $A$ is a normalization factor.

Generally, self energy of an atom $C(z)$ includes self energy of its nucleus $C^{(\mathrm{N})}(z)$ and self energy of its electrons $C^{(\mathrm{e})}(z)$ and can be represented roughly as their sum:

$$C(z) \approx C^{(\mathrm{e})}(z) + C^{(\mathrm{N})}(z - Zm_e - \varepsilon). \tag{8}$$

Here $m_\mathrm{e}$ is electron rest energy, and $\varepsilon$ is electronic binding energy. Consider the electronic self energy function first. Its real part in the leading order represents the Lamb shift of electronic states, which is assumed here to be included into the atomic energies $E_\nu$ ($\nu = i, f$). The imaginary part determines the decay width of a state. For the excited state $|i\rangle$ it is equal up to a factor to the decay width $W_i$ of this state, whereas for the final state (which we assume to be stable) it equals zero:

$$C_i^{(\mathrm{e})}(z) = \langle i|C^{(\mathrm{e})}(z)|i\rangle = -\mathrm{i}\frac{W_i}{2}, \tag{9}$$

$$C_f^{(\mathrm{e})}(z) = \langle f; \mathbf{k}, \varepsilon_\lambda|C^{(\mathrm{e})}(z)|f; \mathbf{k}, \varepsilon_\lambda\rangle = 0. \tag{10}$$

The self energy function of the nucleus, in turn, can be expanded in a Taylor series near the point of its rest energy $E_{\mathrm{N}}$:

$$C_\nu^{(\mathrm{N})}(z) = C_\nu^{(\mathrm{N})}(E_{\mathrm{N}}) + \left.\frac{dC_\nu^{(\mathrm{N})}(z)}{dz}\right|_{z=E_{\mathrm{N}}}(z - E_{\mathrm{N}}), \quad \nu = (i, f). \tag{11}$$

The constant part of this expansion includes the ordinary shift of nuclear levels $\Delta E_{\mathrm{N}}$ caused by vacuum polarization effects and their decay width $\Gamma_0$. As before, the energy shift is assumed to be accounted for in the atomic energy $E_\nu$. Moreover, as the effect of electronic states on the state of the nucleus is negligible, this term is assumed to be the same for both atomic states: $C_i^{(\mathrm{N})}(E_i) = C_f^{(\mathrm{N})}(E_f) = -\mathrm{i}\Gamma_0/2$. The derivative in the second term of (11) taken at the point $z = E_{\mathrm{N}}$ is generally a complex constant. In this work we will restrict ourselves to the case when it is purely real:

$$\left.\frac{dC_\nu^{(\mathrm{N})}(z)}{dz}\right|_{z=E_{\mathrm{N}}} = \rho, \qquad \operatorname{Im}\rho = 0. \tag{12}$$

This derivative is defined by the generalized dynamical equation applied to the problem [10]. This equation contains the nuclear interaction operator and hence, the constant $\rho$ carries information about this interaction. Substituting expressions (9,10) and (11) into the matrix elements of operators (8), we obtain:

$$C_i(z) = -\mathrm{i}\frac{W_i}{2} - \mathrm{i}\frac{\Gamma_0}{2} + \rho(z - E_i), \tag{13}$$

$$C_f(z) = -\mathrm{i}\frac{\Gamma_0}{2} + \rho(z - E_f). \tag{14}$$

Here it is taken into account that

$$E_{\mathrm{N}} + Zm_{\mathrm{e}} + \varepsilon_\nu = E_\nu. \tag{15}$$

The total energy of the decay products is a sum of their rest energies and their total kinetic energy $p^2/2\mu$:

$$E_{dp} = m_1 + m_2 + \frac{p^2}{2\mu}, \tag{16}$$

where $\mu = (m_1 m_2)/(m_1 + m_2)$ and $p = |\mathbf{p}|$. The rest energy of the $k$-th decay product ($k = 1, 2$) is the sum of the rest energy of its nucleus $E_k$, the rest energy of its electrons and their binding energy $\varepsilon_k$:

$$m_k = E_k + Z_k m_{\mathrm{e}} + \varepsilon_k. \tag{17}$$

Using this, we can write for the absolute square of matrix element (7):

$$|S_{fi}|^2 = A\omega\frac{|M_{N\mu}|^2}{(\frac{p^2}{2\mu} - Q - \delta\varepsilon)^2 + \frac{1}{(1-\rho)^2}\frac{\Gamma_0^2}{4}} \times \frac{|M_{fi}|^2}{(\frac{p^2}{2\mu} - Q - \delta\varepsilon + \omega - \omega_{fi})^2 + \frac{1}{(1-\rho)^2}\left(\frac{\Gamma_0}{2} + \frac{W_i}{2}\right)^2}, \tag{18}$$

where $Q = E_{\mathrm{N}} - E_1 - E_2$ is the $Q$-value of the reaction, $\omega_{fi} = \varepsilon_i - \varepsilon_f$ is the atomic transition energy and $\delta\varepsilon = \varepsilon_f - (\varepsilon_1 + \varepsilon_2)$. Let us estimate the quantities which enter this expression. An optical transition frequency has the order of several eV. The width of an atomic level has the order of $10^{-7}$ eV. The magnitude of $Q$-value has in average the order of tens of MeV. Moreover, in this work we consider the nuclei whose decay widths $\Gamma_0$ are comparable to the atomic decay width: $\Gamma_0 \approx W_i$. Summarizing this, one can write: $\Gamma_0 \approx W_i \ll \omega \ll Q$. And, hence, a small parameter $\omega/Q \sim 10^{-6}$ can be introduced.

In the approximation that $|M_{N\mu}|^2$ and $|M_{fi}|^2$ are not dependent on $p$ the integral with respect to $p$ in (1) can be easily taken. Omitting an inessential constant factor and neglecting all non-zero orders of $\omega/Q$, we obtain for probability:

$$\frac{dW}{d\omega} \sim \frac{1}{(1-\rho)^3}\frac{\Gamma_0 + W_i/2}{\Gamma_0/2(\Gamma_0/2 + W_i/2)} \cdot \frac{1}{(\omega - \omega_{fi})^2 + \frac{(2\Gamma_0 + W_i)^2}{4(1-\rho)^2}}. \tag{19}$$

This is a pure Breit-Wigner distribution with the width $W$ defined by the following expression:

$$W = \frac{2\Gamma_0 + W_i}{1-\rho}.$$

As can be seen, from the last formula, the width of the contour represents the sum of the initial state width $(W_i + \Gamma_0)$ and the final state width $(\Gamma_0)$, multiplied by the factor $\rho$ introduced in (11). Small variation of this parameter leads to an essential change in the width of the spectral line. When $\rho = 0$, just the sum of the initial and final state widths remains.

## 3. Conclusion

The self energy function of the radioactive nucleus has been regarded. It is shown to be dependent on a parameter, characterizing the dynamics of nuclear decay. An equation for a spectral line contour of an atom with a radioactive nucleus is obtained. For optical transitions, when the transition frequency is much smaller than the mass excess of the nucleus, this contour is shown to be purely of the Breit-Wigner form. The width of the contour represents a nonlinear combination of the nuclear and atomic decay widths multiplied by a factor containing the above arbitrary constant. This parameter leads to an essential broadening of the spectral line. Physically, this parameter carries information about the nuclear interaction, and hence, definition of this parameter from optical spectra would help to improve our models of nuclear interaction.

## Acknowledgments

This work was funded by the subsidy allocated to Kazan federal University for the state assignment in the sphere of scientific activities.

## References

[1] Barzakh A E, Chubukov I Ya, Fedorov D V, Panteleev V N, Seliverstov M D and Volkov Yu M 2000 *Phys. Rev.* C **61** 034304
[2] Karaivanov D V, Zemlyanoi S G and Myshinsky G V 2005 *Hyperfine Interactions* **162** 195
[3] Feynman R P and Hibbs A R 1965 *Quantum Mechanics and Path Integrals* (New York: McGraw-Hill)
[4] Gainutdinov R Kh 1989 *J. Phys.* A **22** 269-286
[5] Gainutdinov R Kh 1999 *J. Phys.* A **32** 5657-77
[6] Gainutdinov R Kh, Mutygullina A A and Scheid W 2002 *Phys. Lett.* A **306** 1
[7] Gainutdinov R Kh and Mutygullina A A 2002 *Phys. Rev.* C **66** 014006
[8] Gainutdinov R Kh and Moutygoullina A A 1997 *Phys. Atom. Nucl.* **60** 841-847
[9] Gainutdinov R Kh 1995 *ZETP* **108** 1600
[10] Gainutdinov R Kh, Mutygullina A A and Petrova A S 2014 *J. Phys.: Conf. Ser.* **613** 012003

Journal of Physics: Conference Series **714** (2016) 012009 doi:10.1088/1742-6596/714/1/012009

# Self-energy shift of the energy levels of atomic hydrogen in photonic crystal medium

**R Kh Gainutdinov[1], M A Khamadeev[1], O V Steryakov[1], K A Ziyatdinova[1] and M Kh Salakhov[1,2]**

[1] Kazan Federal University, Institute of Physics, 420008, Kremlevskaya St, 18, Kazan, Russian Federation

[2] Tatarstan Academy of Sciences, 420111, Bauman St, 20, Kazan, Russian Federation

E-mail: `Renat.Gainutdinov@kpfu.ru`

**Abstract.** Corrections to the average kinetic energy of atomic electrons caused by the change in electron mass in the photonic crystal medium are investigated. Corresponding shift of energy levels of atoms placed in a photonic crystal is shown to be of order of the ordinary Lamb shift.

The investigation of photonic crystals (PC) is an actual problem today. The main feature of such media is the periodic modulation of dielectric constant along one, two or three directions of space. It yields to the range of forbidden frequencies in a material called photonic band-gaps (PBG). Unusual optical properties of the PC medium stimulate a wide range of its practical applications [1]. Another interesting field of study is quantum electrodynamics (QED) in PC medium. It has been shown that modification of photon density of states (DOS) in PC's lead to the strong emitter-photon coupling [2–4], coherent control of spontaneous emission [5–7], appearance of photon-atom bound states [2, 8, 9] etc. Special attention was given to the modification of the Lamb shift [2,8,10–14] in atoms placed in the photonic crystal medium. By definition the Lamb shift in atoms is caused by the QED corrections to the Coulumb interaction between electrons and the atomic nucleus (see figure 1a) that gives the leading order contribution to the energy of atomic states. This interaction is modified by the processes in which the virtual photons come into play (see figure 1b). The contribution to atomic energy levels from these processes is just the Lamb shift. In principle one should account for the processes of the self interaction of the atomic electrons shown in figure 1c. In the case of atoms in free space these processes contribute only to the electron mass and are subtracted by the counter-terms of the renormalization. However, as it has been shown in Ref. [15], a modification of the of the interaction of the electron with its own radiation field in a PC gives rise to the fact that contributions from the processes in a PC shown in figure 1c differ profoundly from those in free space, and, as a result, these self-energy processes can play an important role. First of all, they change the mass of an electron in the PC medium [15]. In addition they contribute to the kinetic energy of atomic electrons in the PC. The aim of this paper is to investigate this correction to the energy levels of atoms in the PC medium.

The change of the value of the kinetic energy of an electron $\delta E$ is the difference between its values in the PC medium and in vacuum:

$$\delta E^{pc} = E^{pc} - E^{vac}.$$

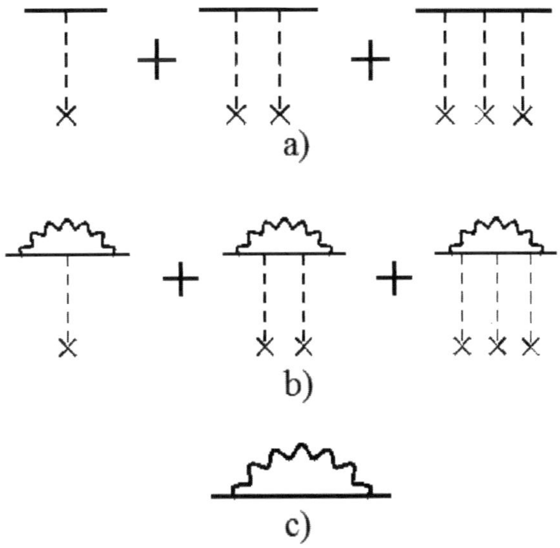


**Figure 1.** a) Processes of the Coulumb interaction between the nucleus and atomic electrons that give the leading order contribution to the energy of atomic states. b) Processes in which between emission and reabsorbtion of a photon a bound electron interacts with the Coulumb field of the nucleus. Just these processes give rise to the Lamb shift. c) Process in which an atomic electron emits and reabsorbs a photon virtually without any interaction with the nucleus. This process changes the electron mass $m_e \to m_e + \delta m_e$ and the kinetic energy $\frac{\mathbf{p}^2}{2m_e} \to \frac{\mathbf{p}^2}{2m_e} - \frac{\mathbf{p}^2}{2m_e}\frac{\delta m_e}{m_e}$.

The one-loop low-energy contribution to the kinetic energy of an electron propagating with momentum $\mathbf{p}$ in the PC medium and in the free space in the Coulomb gauge takes the form (here and below, we use the natural system of units in which $\hbar = c = \varepsilon_0 = 1$)

$$E^{pc} = \sum_{\mathbf{p}'}\sum_{\mathbf{k}n} \frac{\langle \mathbf{p}| H_I^{pc} |\mathbf{p}'; \mathbf{k}, n\rangle \langle \mathbf{p}'; \mathbf{k}, n| H_I^{pc} |\mathbf{p}\rangle}{\frac{\mathbf{p}^2}{2m_e} - \frac{\mathbf{p}'^2}{2m_e} - \omega_{\mathbf{k}n}}$$

and

$$E^{vac} = \sum_{\mathbf{p}'}\sum_{\mathbf{k}\varepsilon_\lambda} \frac{\langle \mathbf{p}| H_I |\mathbf{p}'; \mathbf{k}, \varepsilon_\lambda\rangle \langle \mathbf{p}'; \mathbf{k}, \varepsilon_\lambda| H_I |\mathbf{p}\rangle}{\frac{\mathbf{p}^2}{2m_e} - \frac{\mathbf{p}'^2}{2m_e} - |\mathbf{k}|}$$

correspondingly, where

$$\langle \mathbf{p}'; \mathbf{k}, n| H_I^{pc} |\mathbf{p}\rangle = -\frac{e}{m_e}\frac{1}{\sqrt{V\omega_{\mathbf{k}n}}} \sum_{\mathbf{G}} \mathbf{p} \cdot \mathbf{E}^*_{\mathbf{k}n}(\mathbf{G}) \delta_{\mathbf{p},\mathbf{p}'+\mathbf{k}+\mathbf{G}}$$

and

$$\langle \mathbf{p}| H_I^{pc} |\mathbf{p}'; \mathbf{k}, n\rangle = -\frac{e}{m_e}\frac{1}{\sqrt{V\omega_{\mathbf{k}n}}} \sum_{\mathbf{G}} \mathbf{p} \cdot \mathbf{E}_{\mathbf{k}n}(\mathbf{G}) \delta_{\mathbf{p},\mathbf{p}'+\mathbf{k}+\mathbf{G}},$$

with $\mathbf{E}_{\mathbf{k}n}(\mathbf{G})$ being the coefficients in the plane-wave expansion $\mathbf{E}_{\mathbf{k}n}(\mathbf{r}) = \sum_{\mathbf{G}} \mathbf{E}_{\mathbf{k}n}(\mathbf{G}) e^{i(\mathbf{k}+\mathbf{G})\cdot\mathbf{r}}$ of the Bloch eigenfunctions $\mathbf{E}_{\mathbf{k}n}(\mathbf{r})$ and having the polarization structure $\mathbf{E}_{\mathbf{k}n}(\mathbf{G}) = e^{\mathbf{G}1}_{\mathbf{k}n}\varepsilon_{\mathbf{G}1} + e^{\mathbf{G}2}_{\mathbf{k}n}\varepsilon_{\mathbf{G}2} + e^{\mathbf{G}3}_{\mathbf{k}n}\varepsilon_{\mathbf{G}3}$. Here $\mathbf{G}$ is the reciprocal lattice vector of the photonic crystal ($\mathbf{G} = N_1\mathbf{b}_1 + N_2\mathbf{b}_2 + N_3\mathbf{b}_3$, where $\mathbf{b}_i$ are primitive basis vectors of the reciprocal lattice) [16]. Thus, the PC medium correction to the kinetic energy of a free electron is determined by the expression

$$\delta E^{pc} = -\frac{\alpha}{2m_e^2\pi^2}\left[\sum_n \int\limits_{FBZ} \frac{d^3k}{\omega_{\mathbf{k}n}^2} \sum_{\mathbf{G},\lambda} \left|e_{\mathbf{k}n}^{\mathbf{G}\lambda}\right|^2 \left|\mathbf{p}\cdot\varepsilon_{\mathbf{G}\lambda}\right|^2 - \int \frac{d^3k}{2\mathbf{k}^2}\sum_{\lambda=1}^{2}\left|\mathbf{p}\cdot\varepsilon_\lambda(\mathbf{k})\right|^2\right].$$

If an electron is in the state $|\Psi_n\rangle$, the correction to its energy is given by

$$\delta E_n^{pc} = \langle\Psi_n|\,\delta E^{pc}\,|\Psi_n\rangle\,. \tag{1}$$

Let us consider the example of atomic hydrogen. In this case energy shift (1) takes the form

$$\delta E_a^{pc} = \int d^3p\Psi_a^*(\mathbf{p})\delta E^{pc}(\mathbf{p}/\left|\mathbf{p}\right|)\Psi_a(\mathbf{p}), \tag{2}$$

where $|a\rangle = |njlm\rangle$. Assuming that the modification of the interaction of an electron with its own radiation field in PC medium has not a significant effect on the wave function, the electron wave function can be represented in the form $\Psi_{njlm}(\mathbf{p}) = \frac{1}{p}G_{nl}(p)Y_{lm}(\Omega)$, where $G_{nl}(p)$ is the radial function and $Y_{lm}(\Omega)$ is the spherical function. Taking this into account, (2) can be rewritten

$$\delta E_a^{pc} = -\frac{\alpha\left\langle p^2\right\rangle_a}{2m_e^2\pi^2}\left[\sum_n \int\limits_{FBZ} \frac{d^3k}{\omega_{\mathbf{k}n}^2} \sum_{\mathbf{G},\lambda} \left|e_{\mathbf{k}n}^{\mathbf{G}\lambda}\right|^2 \int d\Omega Y_{lm}^*(\Omega)\left|\frac{\mathbf{p}}{p}\cdot\varepsilon_{\mathbf{G}\lambda}\right|^2 Y_{lm}(\Omega) - \frac{4\pi}{3}\int\limits_0^\infty dk\right], \tag{3}$$

where

$$\left\langle p^2\right\rangle_a = \int\limits_0^\infty dpp^2G_{lm}^*(p)G_{lm}(p).$$

Thus, equation (3) allows us to calculate the self-energy shift of the energy levels of atomic hydrogen if we know the coefficients in the plane-wave expansion of the Bloch eigenfunctions $\mathbf{E}_{\mathbf{k}n}(\mathbf{r})$. Let us derive the value of this shift for the ground state $|1s\rangle$. The corresponding momentum wave function is

$$\Psi_{1s}(\mathbf{p}) = \frac{1}{p}\sqrt{\frac{2}{\pi\beta^3}}\frac{4p}{\left(1+\left(\frac{p}{\beta}\right)^2\right)^2}\frac{1}{\sqrt{4\pi}},$$

where $\beta = m_e\alpha$ and $\alpha$ is fine structure constant. In this case we can simply integrate in (3) over the solid angles $\Omega$ for each $\mathbf{G}$ and $\lambda$ and it yields

$$\delta E_{1s}^{pc} = -\frac{\alpha^2}{2}\frac{4\alpha}{3\pi}\left[\frac{1}{4\pi}\sum_n \int\limits_{FBZ} \frac{d^3k}{\omega_{\mathbf{k}n}^2} \sum_{\mathbf{G},\lambda}\left|e_{\mathbf{k}n}^{\mathbf{G}\lambda}\right|^2 - \int\limits_0^\infty d\omega\right].$$

Taking this into account we can derive the correction to the atomic states in the same way as it was made in Ref. [15]

$$\delta E_{1s}^{pc} = -\frac{\alpha^2}{2}\frac{4\alpha}{3\pi}\int\limits_0^\infty d\omega\frac{N(\omega)-\omega^2}{\omega^2}, \tag{4}$$

where $N(\omega) = N_{DOS}(\omega)D(\omega)$, $N_{DOS}(\omega)$ is the photon density of states

$$N_{DOS}(\omega) = \frac{1}{4\pi}\sum_n \int\limits_{FBZ} d^3k\delta(\omega-\omega_{\mathbf{k}n})$$

and

$$D(\omega) = \sum_{\mathbf{G}} \left|\mathbf{E}_{\mathbf{k}n}(\mathbf{G})\right|^2_{|\omega_{\mathbf{k}n}=\omega} .$$

Thus, the function $N(\omega)$ is closely associated with the DOS. The DOS in the PC medium differs significantly from that in vacuum and could demonstrate singular behavior, see, for example, figure 2. The basic features of the PC medium such as photonic band gap and optical density

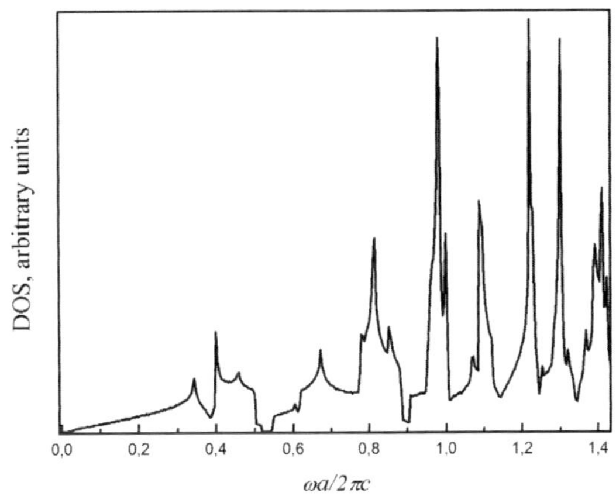


**Figure 2.** Two-dimensional DOS for a triangular lattice of inversion rods in material with $\epsilon = 12$. The radius of an air cylinder is half of the period of the lattice $a$.

could be recaptured by the following model function

$$N(\omega) = \omega^2 n_{eff}^3 \left[1 - h \exp\left(-\frac{(\omega - \omega_0)^2}{\sigma^2}\right)\right] F(\omega), \tag{5}$$

where the factor $F(\omega) = n_{eff}^{-3} + (1 - n_{eff}^{-3})/(\exp\{(\omega - \mu)/\tau\} + 1)$ with $n_{eff} \equiv \sqrt{\bar{\varepsilon}}$ and $\bar{\varepsilon}$ being average dielectric constant. Function $F(\omega)$ allows one to take into account that at high enough photon energies $N(\omega)$ must approach the free space DOS. With this model for the parameters $n_{eff} = 3$, $\mu = 15$ eV, $\tau = 0.01$ eV, $\omega_0 = 1$ eV, $h_0 = 0.96$ and $\sigma_0 = 0.07$ eV our calculations of the self-energy shift of the 1$s$ level of atomic hydrogen have given $\delta E_{1s}^{pc} = -2.2 \cdot 10^{-5}$ eV that is of order of the Lamb shift of 1$s$-state in vacuum. It can be shown that self-energy correction (3) decreases with $n$ as $1/n^2$. As a consequence, the energy levels of Rydberg atomic states should crucially depend on the shifts caused by the change in electrons kinetic energy. This allows one to drive different resonant processes, that could be very important for the optical properties of Rydberg atoms [17].

In conclusion, we have investigated the corrections to the average kinetic energy of atomic electrons caused by the change in electron mass in the photonc crystal medium. These corrections are shown to decrease with $n$ as $1/n^2$ whereas the Lamb shift decreases as $1/n^3$. This fact means that at large values of $n$ the shift caused by the change in electron mass is the main contribution to the correction to energies of atomic states.

**Acknowledgments**
This work was funded by the subsidy allocated to Kazan Federal University for the state assignment in the sphere of scientific activities.

## References

[1] Joannopoulos J D, Johnson S G, Winn J N and Meade R D 2008 *Photonic Crystals: Molding the Flow of Light* (Princeton: Princeton University Press) p 286
[2] John S and Wang J 1990 *Phys. Rev. Lett.* **64** 2418
[3] Zou L *et al* 2015 *J. Lumin.* **158** 281
[4] Kedia S, Sinha S 2015 *J. Phys. Chem.* C **119** 8924
[5] Quang T, Woldeyohannes M, John S and Agarwal G S 1997 *Phys. Rev. Lett.* **79** 5238
[6] Roy C 2010 *J. Phys.* B **43** 235502
[7] Zhu Y S *et al* 2012 *Appl. Phys. Lett.* **100** 081104
[8] John S and Wang J 1991 *Phys. Rev.* B **43** 12772
[9] Bay S, Lambropoulos P and Mölmer K 1997 *Phys. Rev. Lett.* **79** 2654
[10] Zhu S Y, Yang Y, Chen H, Zheng H and Zubairy M S 2000 *Phys. Rev. Lett.* **84** 2136
[11] Vats N, John S and Busch K 2002 *Phys. Rev.* A **65** 043808
[12] Li Z Y and Xia Y 2001 *Phys. Rev.* B **63** 121305(R)
[13] Wang X-H, Kivshar Y S and Gu B-Y 2004 *Phys. Rev. Lett.* **93** 073901
[14] Wang X-H, Kivshar Y S and Gu B-Y 2005 *Science and Technology of Advanced Materials* **6** 814
[15] Gainutdinov R Kh, Khamadeev M A and Salakhov M Kh 2012 *Phys. Rev.* A **85** 053836
[16] Sakoda K 2001 *Optical Properties of Photonic Crystals* (Berlin: Springer) p 232
[17] Nussenzveig P *et al* 1993 *Phys. Rev.* A **48** 3991

Journal of Physics: Conference Series **714** (2016) 012010 doi:10.1088/1742-6596/714/1/012010

# Improvement of near-field enhancement with a grating-assisted gold tapered nanoantenna

**A R Gazizov[1,2], M Zohrabi[1], S S Kharintsev[1] and M Kh Salakhov[1,2]**

[1] Kazan Federal University, 18 Kremlevskaya str., 420008 Kazan, Russian Federation
[2] Academy of Science of the Republic Tatarstan, 20 Bauman str., 420111 Kazan, Russian Federation

E-mail: Almaz.Gazizov@kpfu.ru

**Abstract**. This work is dedicated to the improvement of the near-field enhancement beneath the tip apex due to delocalized plasmons excitation on a sub-wavelength grating engraved on the tip mesoscopic surface. To study conditions of the maximal enhancement we have performed PSO-based optimization of intensity in search space of two parameters. Those parameters are period of the grating and its position in respect to the apex. The grating-patterned tip is illuminated with the incident light with wavelengths of 400 to 1000 nm in our model. All the simulations of electromagnetic waves scattering on the nanoantenna are based on the finite difference time domain method.

## 1. Introduction

Optical antenna is a mesoscopic convertor of freely propagating modes to localized modes of electromagnetic waves and vice versa [1]. They contribute to the strengthening of the interaction of freely propagating light with a single emitters or a receivers of the nanoscopic light, thus contributing to the formulation of many applications of near-field optics [2]. For example, they can enhance the signal of Raman scattering and fluorescence by several orders in magnitude and localize electromagnetic energy at the nanoscale [3].

Amplification of electric field near the gold tip is the result of surface plasmon resonance, which depends on the geometry of the antenna, with an influence of the effect of a lightning rod [4]. In tip-enhanced near-field optical microscopy (TENOM), laser light illuminates the sample in presence of the tip. In case of direct illumination, detected intensity is composed of the signal from subwavelength area beneath the tip and the background signal from diffraction-limited area. Many studies are aimed to reduce the second part of the signal [5].

Recently, Raschke *et al.* [6] have introduced an alternative method based on the adiabatic focusing of plasmons at the apex of the probe to perform the spectroscopy of tip-enhanced Raman scattering (TERS). The advantage of this method is the formation of a nanoscale Raman excitation source, which contributes to the suppression of the internal background Raman signal. Adiabatic nanofocusing is a suitable way of transforming the optical signal of the far field in the near field of the localized light source [7], and it has a huge potential to perform background-free nanoscale chemical spectroscopy. The investigation of the adiabatic focusing by grating-patterned optical antennas can be found in Refs. [8-10].

In order to reduce the influence of the background signal we consider a gold taper probe with a sub-wavelength grating. When incident light illuminates the grating, it generates surface plasmons, which are moving to the apex of the optical antenna. This leads to compression and localization of electromagnetic energy. In this article, we study the geometric conditions of the grating illumination, under which the maximal intensity occurs near the end of the nanoantenna. We find an optimal value of grating period and its position in respect to the apex (see figure 1, parameters *a*, *b*) by means of the Particle Swarm Optimization (PSO).

## 2. Optimization method

Finite-difference time-domain method (FDTD) is a successful application in an extremely wide range of tasks, such as light scattering from metal objects and dielectrics, antennas, microstrip circuits and electromagnetic absorption in the human body, exposed to radiation [11]. It also uses a different algorithm — particle swarm optimization (PSO). PSO algorithm is the one of the most powerful methods for solving the non-smooth global optimization problems while there are some disadvantages of the PSO algorithm [12].

Kennedy and Eberhart were the first to set the solution of complex nonlinear optimization problems by simulating the behavior of bird flocks. They introduced the concept of objective function of optimization by using a particle swarm [12]:

$$\min_{X \in R^n} \Phi(X) = \Phi(X^*) \tag{1}$$

The particle swarm optimization is a multi-agent parallel search technique that works with a swarm of particles and each particle represents a potential solution in the swarm. All particles move in the multidimensional search space $R^n$, and each particle modifies its position depending on its own experience and that of neighbors. In the PSO technique, all particles are initiated randomly and their values of the objective function are estimated by finding a local best (the best value of each particle) and a global best (the best value in the whole swarm). This is followed by a cycle of the search for optimal value. At first, in the cycle, the velocities of particles are updated according to global and local best; then, the positions of particles are updated according to the velocities of particles. This cycle repeats many times, until the stopping criterion is satisfied [13]. There are two developed PSO algorithms, namely the global best (gbest) and local best (lbest) PSO, which differ by the size of the environment [14].

$$X_{i,t+1} = X_{i,t} + V_{i,t+1} \tag{2}$$

$$V_{i,t+1} = \alpha V_{i,t+1} + U[0,\beta] \otimes (X_{i,t}^{b} - X_{i,t}) + U[0,\gamma] \otimes (X_{g,t} - X_{i,t}) \tag{3}$$

Here U[*a*,*b*] is a r-dimensional vector of pseudorandom numbers uniformly distributed in the interval [*a*,*b*]; ⊗ is the symbol for componentwise multiplication of vectors. $X_{i,t}^{b}$ is vector of coordinates of the particle $P_i$ with the best (by means of (1)) value of the objective function Φ(X) for the time search; $X_{g,t}$ is vector of coordinates neighboring a given particle with the best for the value of the objective function Φ(X); α, β, γ are free parameters of the algorithm.

## 3. Results and discussion

In this paper, we investigated dependence of the field enhancement on the parameters *a* and *b* of the grating. We proposed a gold pointed probe with the curvature radius of the tip 10 nm, the cone angle of 30° and with the grating period, which lies between 300 and 600 nm. We considered the laser beam as a plane wave with a wavelength of 400-1000 nm. The image of the considered model and the simulation results for the grating with accounting period are below. Figure 1 shows a diagram of the proposed optical antenna with lattice period *a* and the distance between the end of the probe and the

Journal of Physics: Conference Series **714** (2016) 012010 doi:10.1088/1742-6596/714/1/012010

grating along the axis of the probe $b$. These two parameters form a two-dimensional search space for PSO.

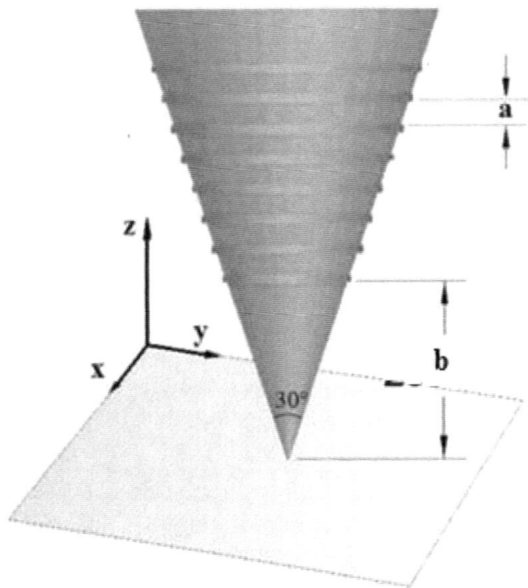


**Figure 1**. Considered geometry for optical antenna.

The objective function is a sum of squared differences between the enhancements of the intensity near the tip apex for a given particle of the swarm and for other particles. One can express the complex enhancement factor of the field as follows:

$$\alpha_{\text{eff},\|} = 2\pi\varepsilon_0\rho^3 f_e \tag{4}$$

Expression (4) for the longitudinal polarizability of the probe occurs from the condition that the field on the surface of the edge is numerically equal to the field that is specified by $f_e E_0$ [15]. Using the formalism of the Green's function one can obtain the expression of the resultant field [15-17]:

$$\mathbf{E}(\mathbf{r},\omega) = \mathbf{E}_0(\mathbf{r},\omega) + \frac{\omega^2}{\varepsilon_0 c^2}\ddot{\mathbf{G}}_0(\mathbf{r},\mathbf{r_0};\omega)\alpha_{\text{eff}}\mathbf{E_0}(\mathbf{r_0},\omega) \tag{5}$$

With the usage of PSO algorithm and FDTD-simulation, we obtained the following results. The optimal values of $a$ and $b$, which correspond to the maximum of intensity, are for $a$=450 nm, and for $b$=1200 nm. We used the algorithm of global best. For the numerical simulation the dielectric function of gold was taken from Ref. [18]

Figure 2 shows the intensity distribution around the tip in the X and the Y-axis versus wavelength with the proposed period of the grating and with the angle of incidence of the laser light equal to 90°.

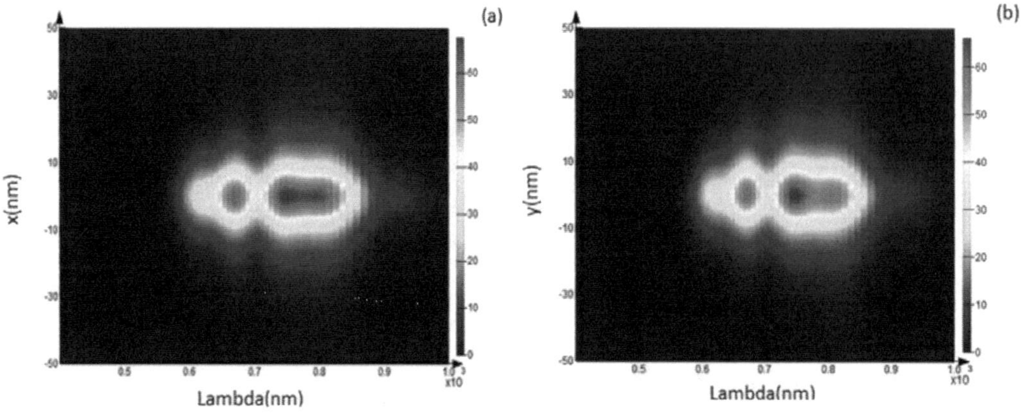


**Figure 2**. a) intensity distribution along the X-axis vs. wavelength at Y=0, Z=0, b) intensity distribution along the Y-axis vs. wavelength at X=0, Z=0.

In case of illumination of the grating by broadband radiation with wavelengths of 400 to 1000 nm, we obtain enhancement of overall intensity of radiation beneath the tip apex. The graphics of dependence of intensity versus $a$ and $b$ are shown in figure 3. We can see that there are some resonance peaks and maximal enhancement of overall intensity by a factor of ~30.

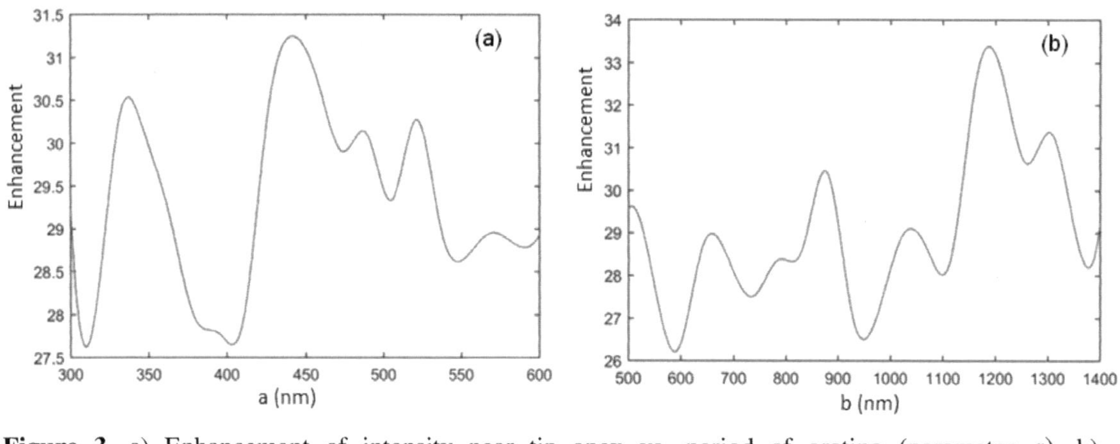


**Figure 3**. a) Enhancement of intensity near tip apex vs. period of grating (parameter $a$). b) Enhancement of intensity near tip apex vs. position of grating in respect to the apex (parameter $b$).

## 4. Conclusion

The results show that the grating-assisted gold tip can compress the electric field by ~8 times and localize it within ~10 nm. It should be kept in mind that with side illumination one can expect a significant suppression of the background signal. Numerical simulations have shown shifts in the resonance band when the grating period changes.

## Acknowledgments

This work was funded by the subsidy allocated to Kazan federal University for the state assignment in the sphere of scientific activities. This work was financially supported by the Russian Foundation for Basic Research (No. 15-42-02339). This work was done using equipment of Federal Center of Shared Equipment of Kazan Federal University.

**References**

[1] Novotny L and van Hulst N 2011 *Nature Photonics* **5** 83-90
[2] Maier S A and Atwater H A 2005 *J. Appl. Phys.* **98** 011101
[3] Novotny L 2011 *Physics today* **64** 47-52
[4] Novotny L and Hecht B 2006 *Principles of Nano-optics* (Cambridge: Cambridge University Press)
[5] Hartschuh A 2008 *Angewandte Chemie International Edition* **47** 8178-8191
[6] Berweger S, Atkin J M, Olmon R L and Raschke M B 2010 *J. Phys. Chem. Lett.* **1** 3427-3432
[7] Stockman M I 2004 *Phys. Rev. Lett.* **93** 137404
[8] Ropers C, Neacsu C C, Raschke M.B, Albrecht M, Lienau C and Elsaesser T 2008 *Japanese Journal of Applied Physics* **47** 6051-6054
[9] Ropers C, Neacsu C C, Elsaesser T, Albrecht M, Raschke M B and Lienau C 2007 *Nano Lett.* **7** 2784-2788
[10] Giugni A, Torre B, Toma1 A, Francardi M, Malerba1 M, Alabastri A, Proietti Zaccaria1 R, Stockman M I and Di Fabrizio E 2013 *Nature Nanotechnology* **8** 845-852
[11] Siriwitpreecha A, Rattanadecho P and Wessapan T 2013 *International Journal of Heat and Mass Transfer* **65** 423-434
[12] Kennedy J and Eberhart R C 1995 *Proceedings of IEEE International conference on Neural Networks* **4** 1942-1948
[13] Taşgetiren M F and Liang Y-C 2004 *Journal of Economic and Social Research* **5** 1-20
[14] Das S, Abraham A and Konar A 2008 *Pattern Recognition Letters* **29** 688-699
[15] Novotny L and Stranick S J 2006 *Annu. Rev. Phys. Chem.* **57** 303-331
[16] Maximiano R V, Beams R, Novotny L, Jorio A and Cançado L G 2012 *Phys. Rev.* B **85** 235434
[17] Rogov A M, Gazizov A R, Kharintsev S S and Salakhov M Kh 2014 *J. Phys.: Conf. Ser.* **560** 012008
[18] Johnson P B and Christy RW 1972 *Phys. Rev.* B **6** 4370-4379

# Longitudinal polarizability and enhancement factor of a tapered optical gold nanoantenna

**A R Gazizov[1,2], S S Kharintsev[1] and M Kh Salakhov[1,2]**

[1] Kazan Federal University, 18 Kremlevskaya str., 420008 Kazan, Russian Federation
[2] Academy of Science of the Republic Tatarstan, 20 Bauman str., 420111 Kazan, Russian Federation

E-mail: Almaz.Gazizov@kpfu.ru

**Abstract**. This work focuses on the mechanism of electric field enhancement near a tapered optical antenna and the calculation of a complex field enhancement factor as a function of tip material, its curvature radius and cone angle. In this paper, an analytical model of longitudinal polarizability, taking into account retardation and dynamic polarization effects, is developed for evaluating the field enhancement factor.

## 1. Introduction

Optical antennas mediate energy between propagating electromagnetic waves and non-propagating near-fields [1]. The antennas are designed for effective concentration of optical field energy below the Abbe's diffraction limit and increasing the efficiency of tip-sample interactions [2]. There are the optical antennas that have found wide application in various microscopic techniques such as tip-enhanced Raman scattering (TERS) [3], antenna-assisted photoelectric cells [4], cancer treatment by the plasmonic probes [5], etc.

Physical properties of a taper optical antenna are characterised by three parameters: radius of curvature $\rho$, cone angle $\alpha$ and dielectric function $\varepsilon(\omega)$ [6]. Of the great importance is the field enhancement factor that depends on spatial distribution of induced currents. Despite the fact, that a lot of research groups worldwide have routinely used experimental tools of near-field microscopy and spectroscopy [7], some theoretical aspects in physics of the field enhancement still remain unsolved. In particular, it concerns a study of the field enhancement factor depending on antenna's parameters [8].

In Ref [9] authors have used a simple numerical model in which the tip apex is substituted for an anisotropic sphere. In this paper, we propose an analytical model of the tip apex, which suits to be used in the infrared region. We calculate the longitudinal polarizability of the nanoantenna in the approximation of the tip apex to be a prolate spheroid. The dynamic polarization and radiative damping of electric oscillations are taken into account by integration of retarded fields of elementary dipoles over the spheroid volume. Analytical results are compared to FDTD-based numerical simulation.

## 2. Results and discussion

Traditionally, the electric field around the tip of the nanoantenna is described with Green's function formalism [10-13]. By replacing the nanoantenna with a metallic nanosphere with a curvature radius of the tip apex $\rho$ and the cone angle $\alpha$ the electric field in a tip-sample system can be written in the following form

$$\mathbf{E}(\mathbf{r},\omega) = \mathbf{E}_0(\mathbf{r},\omega) + \frac{\omega^2}{\varepsilon_0 c^2}\ddot{\mathbf{G}}_0(\mathbf{r},\mathbf{r_0};\omega)\alpha_{\text{eff}}\mathbf{E_0}(\mathbf{r_0},\omega), \tag{1}$$

where $\mathbf{E}_0$ is an incident laser field, $\omega$ is an angular frequency of the incident radiation, $\ddot{\mathbf{G}}_0(\mathbf{r},\mathbf{r_0};\omega)$ is a Green's function in free space and $\alpha_{\text{eff}}$ is an effective polarizability, that contains information on the parameters of the optical antenna.

The effective polarizability of the small metallic sphere $\alpha_{\text{eff}}$ in equation (1) reads as [14]

$$\alpha_{\text{eff}} = 4\pi\varepsilon_0\rho^3 \begin{pmatrix} \frac{\varepsilon_m - 1}{\varepsilon_m + 2} & 0 & 0 \\ 0 & \frac{\varepsilon_m - 1}{\varepsilon_m + 2} & 0 \\ 0 & 0 & f_e/2 \end{pmatrix} \tag{2}$$

$$\alpha_{\text{eff},\perp} = 4\pi\varepsilon_0\rho^3 \frac{\varepsilon_m - 1}{\varepsilon_m + 2} \tag{2a}$$

$$\alpha_{\text{eff},\|} = 2\pi\varepsilon_0\rho^3 f_e \tag{2b}$$

where $f_e$ is a complex field enhancement factor, $\varepsilon_m$ is a dielectric function of the sphere. The expression for $\alpha_{\text{eff},\|}$ originates from the requirement that the magnitude of the field at the surface of the tip is equal to the computationally calculated field set equal to $f_e E_0$ [15].

Modeling and simulating the complex field enhancement factor at various wavelengths are shown in figure 1. The parameters of the tapered tip are $\rho = 10$ nm, $\alpha = 30°$. For the numerical simulation the dielectric function of gold is approximated from Ref. [16] with a modified Drude model with a permittivity at infinite frequency of $\varepsilon_\infty = 7.5$, plasma frequency of $\omega_p = 1.38 \cdot 10^{16}$ rad/s and relaxation time of $\tau = 30$ fs.

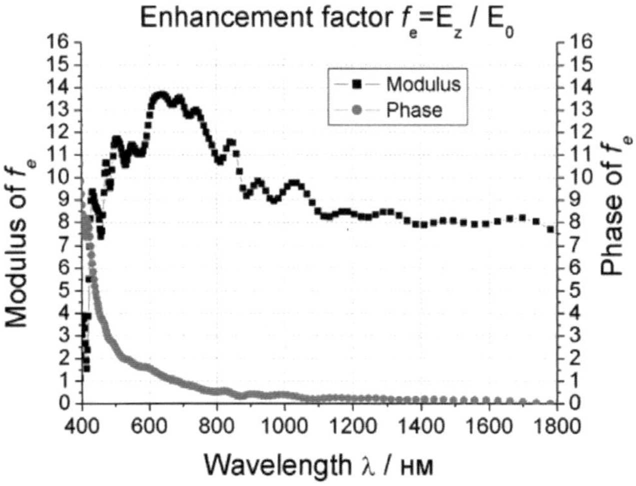


**Figure 1**. The computationally simulated enhancement factor modulus and phase for the gold tapered nanoantenna.

Since the simple sphere model does not provide anisotropy of the polarizability of the tapered optical antenna, we consider prolate spheroid as the model of the tip. Longitudinal polarizability of the metallic spheroidal particle in quasi-static approximation can be written as [17]:

$$\alpha_{\|} = \varepsilon_0 V \frac{\varepsilon_m - 1}{1 + n_{\|}(\varepsilon_m - 1)}, \tag{3}$$

where $n_{\|}$ is a geometric depolarization factor of ellipsoid. The depolarization factor is determined by the following expression for the prolate spheroid [17]:

$$n_{\|} = \frac{1-e^2}{e^2}\left(\frac{1}{2e}\ln\frac{1+e}{1-e} - 1\right), \tag{4}$$

where $e$ is the eccentricity of the ellipse.

The quasi-static polarizability does not assume retardation processes such as dynamic polarization and self-radiation reaction in the taper nanoantenna, so the polarizability in equation (3) is not correct enough. To take these effects into account we use an idea that the electric field $\mathbf{E}$ inside the gold tip is a combination of the incident field $\mathbf{E}_0$ and depolarization field $\mathbf{E}_{dep}$. The polarization vector $\mathbf{P}$ can be expressed in these terms as [18]:

$$\mathbf{P} = \varepsilon_0(\varepsilon - 1)(\mathbf{E}_0 + \mathbf{E}_{dep}). \tag{5}$$

The depolarization field is a sum of all the retarded dipolar fields $\mathbf{dE}_{dep}$ generated by elementary dipoles $\mathbf{dp}(\mathbf{r}) = \mathbf{P}\mathrm{d}V(\mathbf{r})$ of each volume element $\mathrm{d}V$ inside the volume of the ellipsoid. Let the origin of spherical coordinate system be located at one of the foci of the ellipsoid and Z-axis be aligned with its major axis. To find the depolarization field we should integrate the Z-component of the field $\mathbf{dE}_{dep,\|}$ generated at the origin by elementary dipole $\mathrm{d}p_{\|}(r)$ [18]:

$$\mathrm{d}E_{dep,\|} = \frac{1}{4\pi\varepsilon_0}\left[\frac{1}{r^3}(3\cos^2\theta - 1) + \frac{k^2}{2r}(\cos^2\theta + 1) + \mathrm{i}\frac{2}{3}k^3\right]\mathrm{d}p_{\|}(r), \tag{6}$$

where the retarded dipolar field generated by $\mathrm{d}p_{\|}(r)$ has been expanded into a power-series of $(kr)$ up to the power $k^3$ [19]. Integration over the volume of prolate spheroid yields:

$$E_{\mathrm{dep}} = \frac{1}{4\pi\varepsilon_0}\left(-4\pi n_{\|} + 2\pi(1-n_{\|})k^2b^2 + \mathrm{i}\frac{8\pi}{9}k^3ab^2\right)P. \tag{7}$$

Here, $a$ and $b$ is a major and a minor semi-axis of prolate spheroid respectively. Inserting this result into equation (5) and solving it for $P$, one obtains

$$P_{\|} = \frac{\varepsilon - 1}{1 + (\varepsilon-1)n_{\|} + (\varepsilon-1)\frac{n_{\|}-1}{2}k^2b^2 - (\varepsilon-1)\mathrm{i}\frac{2}{9}k^3ab^2}\varepsilon_0 E_0, \tag{8}$$

$$\alpha_{\|} = \frac{\varepsilon - 1}{1 + (\varepsilon-1)n_{\|} + (\varepsilon-1)\frac{n_{\|}-1}{2}k^2b^2 - (\varepsilon-1)\mathrm{i}\frac{2}{9}k^3ab^2}\varepsilon_0 V. \tag{9}$$

We have obtained this result by using of an approximation of homogeneous field and polarization. As we see this polarizability in equation (9) differs from the polarizability of the quasi-static case in equation (3) by the denominator. The term proportional to $k^2$ is referred to the dynamic polarization, because this term vanishes in the case of $k = 0$ (static field). It also contributes only to the real part of the polarizability that corresponds to a change of effective depolarization factor of the spheroidal particle. This effect occurs due to the non-locality of the scattering process in time. The incident electromagnetic wave reaches different parts of the scatterer at different moments, so induced elementary dipoles are oscillating with different phases. Another reason for this term in the denominator is the excitation of electron plasma waves, which bound the field at the surface of the particle.

The term proportional to $k^3$ in the denominator is the radiation-damping correction of the quasi-static polarizability. It occurs due to the accelerated motion of charges inside the volume of the nanoantenna. It accounts for damping of the dipole by radiative losses and results in broadening and the reduction in intensity of the resonance band for the large particle. Since the antenna is a mediator between propagating electromagnetic waves and inductive localized field, one cannot consider it as a small volume particle and the analysis must contain the contribution of these two effects.

If we correct the result with the account of the retardation effect of the incident field analogous to the case of polarization of the sphere according to a power-series expansion of the Mie coefficients ${}^{\mathrm{e}}B_1$ [18], for the complex enhancement factor of the field, we obtain:

$$f_{e,\|} = \frac{2}{3}\frac{1}{1-e^2}\frac{(\varepsilon_{\mathrm{m}}-1)(1-k^2p^2/10)}{1 + (\varepsilon_{\mathrm{m}}-1)n_{\|} + (0.7\varepsilon_{\mathrm{m}}-1)\frac{n_{\|}-1}{2}k^2b^2 - (\varepsilon_{\mathrm{m}}-1)\mathrm{i}\frac{2}{9}k^3ab^2} \tag{10}$$

here $p$ is the ellipse parameter equal to the curvature radius of the tip apex $p = a(1-e^2)$. The eccentricity of the ellipse expressed in terms of the cone angle reads as $e = \sqrt{1-\tan^2\alpha}$.

Figure 2 shows a good agreement with the numerical simulation of the enhancement factor of the tapered antenna in the near-infrared region. Differences between both factors occur in the resonance band of the spheroidal particle. The tapered nanoantenna is a semi-infinite object and only its tip influences the polarizability. This means that surface plasmons excited at the tip apex are not strongly localized plasmons. This leads to a diffuse resonant band. In contrast, the spheroidal particle has a well-localized plasmon modes and a distinct resonance band. Hence, one cannot use this model to describe the resonance effects. The neglected processes become appreciable at shorter wavelengths and the small particle approximation of the tip becomes incorrect. Nevertheless, resonance effects vanish outside the resonance band at longer wavelengths and the small particle approximation remains valid.

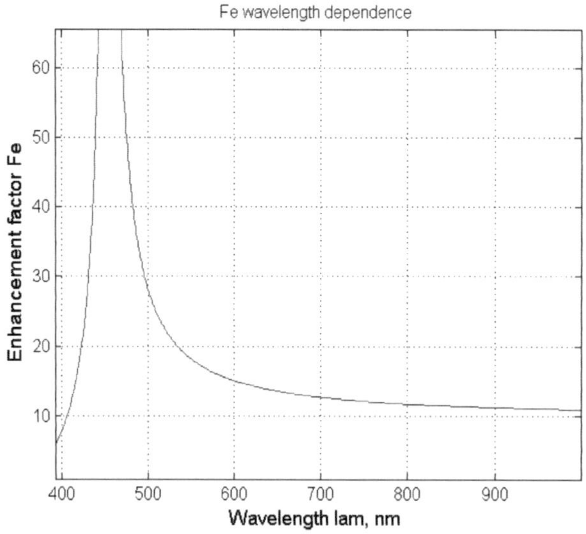


**Figure 2**. The absolute value of the enhancement factor of a gold spheroidal particle. The dielectric function of gold was taken from the modified Drude model with same parameters.

## 3. Conclusion

We have simulated the complex enhancement factor of the field for the gold tapered optical antenna at various wavelengths. We have proposed a prolate spheroid model to describe electric properties of the tip of the gold nanoantenna in the near infrared. This model accounts retardation of electric field by considering the dynamic polarization effect and radiative damping of electric oscillations in the framework of homogeneous polarization approximation. The model of polarizability remains valid at longer wavelengths.

**Acknowledgments**

Authors are grateful to Prof. A.I. Fishman for helpful discussions on the subject of near-field physics. This work was funded by the subsidy allocated to Kazan federal University for the state assignment in the sphere of scientific activities. This work was financially supported by the Russian Foundation for Basic Research (No. 15-42-02339). This work was done using equipment of Federal Center of Shared Equipment of Kazan Federal University.

## References

[1] Novotny L and van Hulst N 2011 *Nature Photonics* **5** 83-90
[2] Maier S A and Atwater H A 2005 *J. Appl. Phys.* **98** 011101
[3] Georgi C and Hartschuh A 2010 *Appl. Phys. Lett.* **97** 143117
[4] Atwater H A and Polman A 2010 *Nature Mater.* **9** 205-213
[5] Bardhan R *et al* 2011 *Acc. Chem. Res.* **44** 936-946
[6] De Angelis F *et al* 2009 *Nature Nanotechnology* **5** 67-72
[7] Mino T, Saito Y and Verma P 2014 *ACS Nano* **8** 10187-10195

Journal of Physics: Conference Series **714** (2016) 012011 doi:10.1088/1742-6596/714/1/012011

[8] Novotny L 2011 *Physics today* **64** 47-52
[9] Bouhelier A, Beversluis M R and Novotny L 2003 *Appl. Phys. Lett.* **82** 4596-4598
[10] Cançado L G, Jorio A, Ismach A, Joselevich E, Hartschuh A and Novotny L 2009 *Phys. Rev. Lett.* **103** 186101
[11] Maximiano R V, Beams R, Novotny L, Jorio A and Cançado L G 2012 *Phys. Rev.* B **85** 235434
[12] Novotny L and Hecht B 2006 *Principles of Nano-optics* (Cambridge: Cambridge University Press)
[13] Rogov A M, Gazizov A R, Kharintsev S S and Salakhov M Kh 2014 *J. Phys.: Conf. Ser.* **560** 012008
[14] Bouhelier A, Beversluis M, Hartschuh A and Novotny L 2003 *Phys. Rev. Lett.* **90** 013903
[15] Novotny L and Stranick S J 2006 *Annu. Rev. Phys. Chem.* **57** 303-331
[16] Johnson P B and Christy RW 1972 *Phys. Rev.* B **6** 4370-4379
[17] Landau L D and Lifshitz E M 1984 *Electrodynamics of continuous media* (Oxford: Butterworth-Heinemann)
[18] Meier M and Wokaun A 1983 *Optics Letters* **8** 581-583
[19] Born M and Wolf E 1980 *Principles of Optics*, 6th ed. (Oxford: Pergamon)

# Surface electric potential of linear periodic charge density

**A V Koryukin[1,2], A A Akhmadeev[1] and M Kh Salakhov[1,2]**

[1] Kazan Federal University, Institute of Physics, 420008, Kremlevskaya St, 18, Kazan, Russian Federation

[2] Tatarstan Academy of Sciences, 420111, Bauman St, 20, Kazan, Russian Federation

E-mail: Artem.Koryukin@kpfu.ru

**Abstract.** Distribution of the surface electric potential allows visualizing surface plasmons and gives information about parameters of the plasmons. Some plasmonic structures have a periodic structure, for example hybrid self-assembled photonic-plasmonic crystals. Earlier we presented a new method of calculation of the potential for quasiperiodic and periodic structures. In this paper we compare the method with other methods that have different algorithms and formulas.

## 1. Introduction

Photonic-plasmonic crystals (PPCs) [1–7] have attracted increased attention recently. PPC is a hybrid structure with combination of dielectric nanophotonic elements with metallic ones supporting surface plasmon polaritons (SPP) [5]. Photonic crystals are materials with spatial periodicity in their dielectric function (or, equivalently, a periodical index of refraction) [8]. Such periodicity yields a strong modification of the electromagnetic field that manifests itself in new quantum electrodynamical effects [9]. PPC are used for easily coupled whispering gallery plasmons [10], surface-enhanced Raman spectroscopy [11, 12], coherent fluorescence emission [13], etc. The effect of surface plasmon resonance underlies the unusual properties of PPC. SPP are electromagnetic excitations propagating at the interface between a dielectric and a conductor [14]. Surface plasmons are collective oscillations in electron density at the surface of a metal [15]. The plasmons influence the surface electric potential. The potential gives rich information about properties and characteristics of the plasmons. Kelvin Probe Force Microscopy (KPFM) allows to use this advantage. It's a modification of atomic force microscopy, which allows to obtain a map of potential distribution [16]. KPFM is an exceptional method for imaging and characterization of plasmonic modes at optical frequencies [17]. Findings [17] opened the way for using KPFM under optical illumination for direct imaging of SPPs with nanometric spatial resolution and sensitivity, established KPFM as a preferred candidate for experimental characterization of SPP nanodevices.

Using KPFM can give advantages to investigation of PPC. Because of dealing with a periodically structured medium, we can use advantages and methods, which are developed for periodic medium, the Bloch's theorem [18] for example. By applying the theorem to distributions of charge and potential a new method of building the map of electric potential distribution was developed [19]. In this work, we used the method for calculation the map of electric potential distribution for an infinitely long, thin periodically charged thread and compared the method with other methods, that have different algorithms and formulas.

## 2. Results

We take an infinitely long, thin periodically charged thread as an object for comparison of methods for calculation of distribution of the electric potential. The potential can be determined via the Coulomb's law:

$$V(\mathbf{r}) = \int \frac{\rho(\mathbf{r}')}{4\pi\varepsilon_0 \left|\mathbf{r}-\mathbf{r}'\right|} \mathrm{d}^3\mathbf{r}', \tag{1}$$

where $V$ is the potential, $\rho$ is the volume charge density.

Let us write the charge distribution for our object:

$$\rho(x, y, z) = \lambda_0 \cos(x), \tag{2}$$

where $\lambda_0$ is the linear charge density.

Substituting (2) into (1), we obtain:

$$V(\mathbf{x}, \mathbf{y}, \mathbf{z}) = \frac{1}{4\pi\varepsilon_0}\lambda_0 \iint \frac{\cos(x)}{\sqrt{(x-x')^2+(y-y')^2}} dxdy. \tag{3}$$

The calculation of the electric potential, obtained by expression (3) for the infinitely long, thin periodically charged thread, lying along the axis x is presented in figure 1. We used the trapezoidal rule for calculation of the integral.

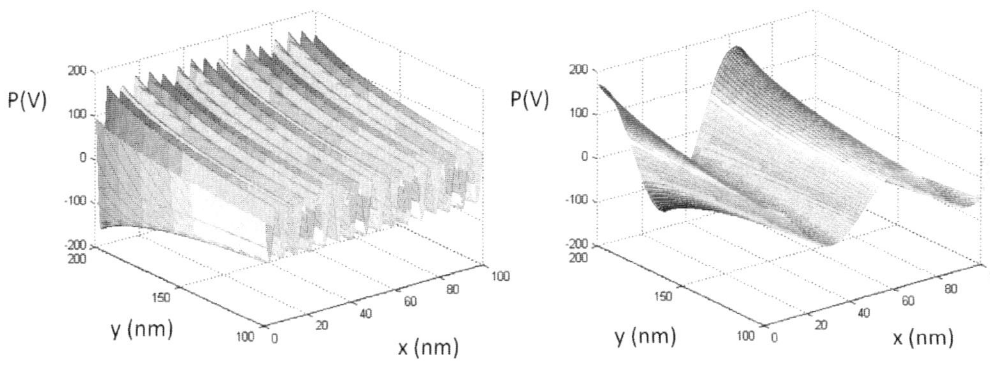


a b

**Figure 1.** Distribution of the electric potential for $\lambda_0 = 10^{-19}$ C, $\frac{1}{4\pi\varepsilon_0} = 10^{10}$ V*m/C. (a), cos(x). (b), cos(0.1*x).

In [19] we presented a method of calculation of the potential for a structure of arbitrary design, but periodic in one direction . We used the Bloch's theorem for periodic medium with the aim of the Coulomb's law simplification. The theorem allowed to get the wave function in periodic repeating medium in the form of a periodic function.

Using these results, we can calculate the map of the potential for the infinitely long, thin periodically charged thread. Let us write the charge distribution:

$$\rho(x', y', z') = \lambda_0 \cos(kx')\delta(y', 0)\delta(z', 0), \tag{4}$$

$$\rho_c(x', y', z') = \lambda_0 e^{ikx'} \delta(y', 0)\delta(z', 0), \tag{5}$$

$$\rho(x', y', z') = Re\rho_c(x', y', z'), \tag{6}$$

where $\mathbf{k}$ is the crystal wave vector.

In [19] we expanded in a series the charge distribution:

$$F_k^{(\rho_c)}(y', z'; G') = \lambda_0 \delta(y', 0)\delta(z', 0)\delta_{0,G'}, \tag{7}$$

$$F_k^{(V)}(y, z; G) = \frac{1}{4\pi\varepsilon_0}\lambda_0 \delta_{0,G} D_k(y', z'; G), \tag{8}$$

where $G$ is the vector of reciprocal lattice, $F_k^{(V)}(y, z; G)$ and $F_k^{(\rho)}(y, z; G)$ are the coefficients of expansion for the potential and charge distribution respectively.

The electric potential can be presented in the form [19]:

$$V_k(x, y, z) = \sum_G F_k^{(V)}(y, z; G)e^{i(k+G)x} = \frac{1}{4\sqrt{2}\pi^2\varepsilon_0}\lambda_0 e^{ikx} D_k(y', z'; 0), \tag{9}$$

where $D_k(y', z'; 0)$ is the integral

$$D_k(y', z'; 0) = \int \frac{e^{ikx''}}{\sqrt{x''^2 + a^2}} dx''. \tag{10}$$

The calculation of the electric potential, obtained by expression (9) for the infinitely long, thin periodically charged thread, lying along the axis x is presented in figure 2. Here we also used the trapezoidal rule for calculation of the integral.

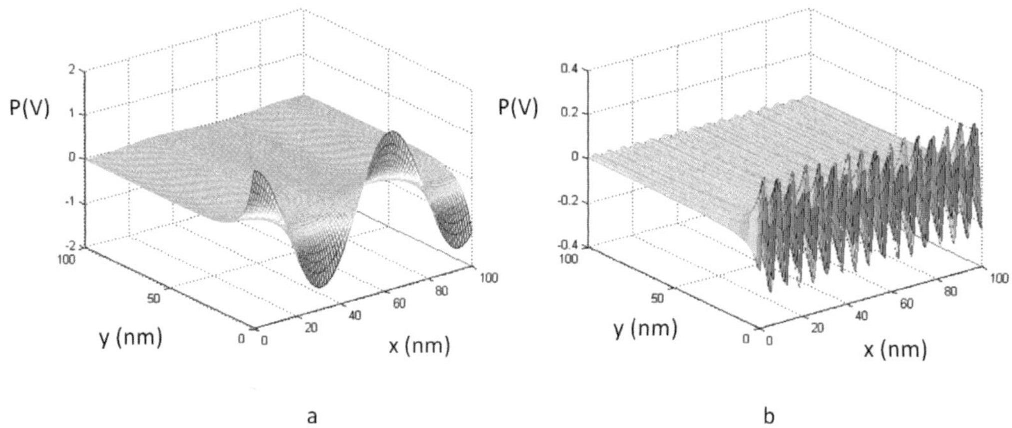


**Figure 2.** Distribution of the electric potential for $\lambda_0 = 10^{-19}$ C, $\frac{1}{4\pi\varepsilon_0} = 10^{10}$ V*m/C. (a), $\mathbf{k}$ = 0.1 $nm^{-1}$. (b), $\mathbf{k}$ = 1 $nm^{-1}$.

Further, we modified the algorithm of calculation of expression (9). We combined two algorithms for calculation of the first and second models. The new distributions is very different from the previous ones, that is shown in figure 3.

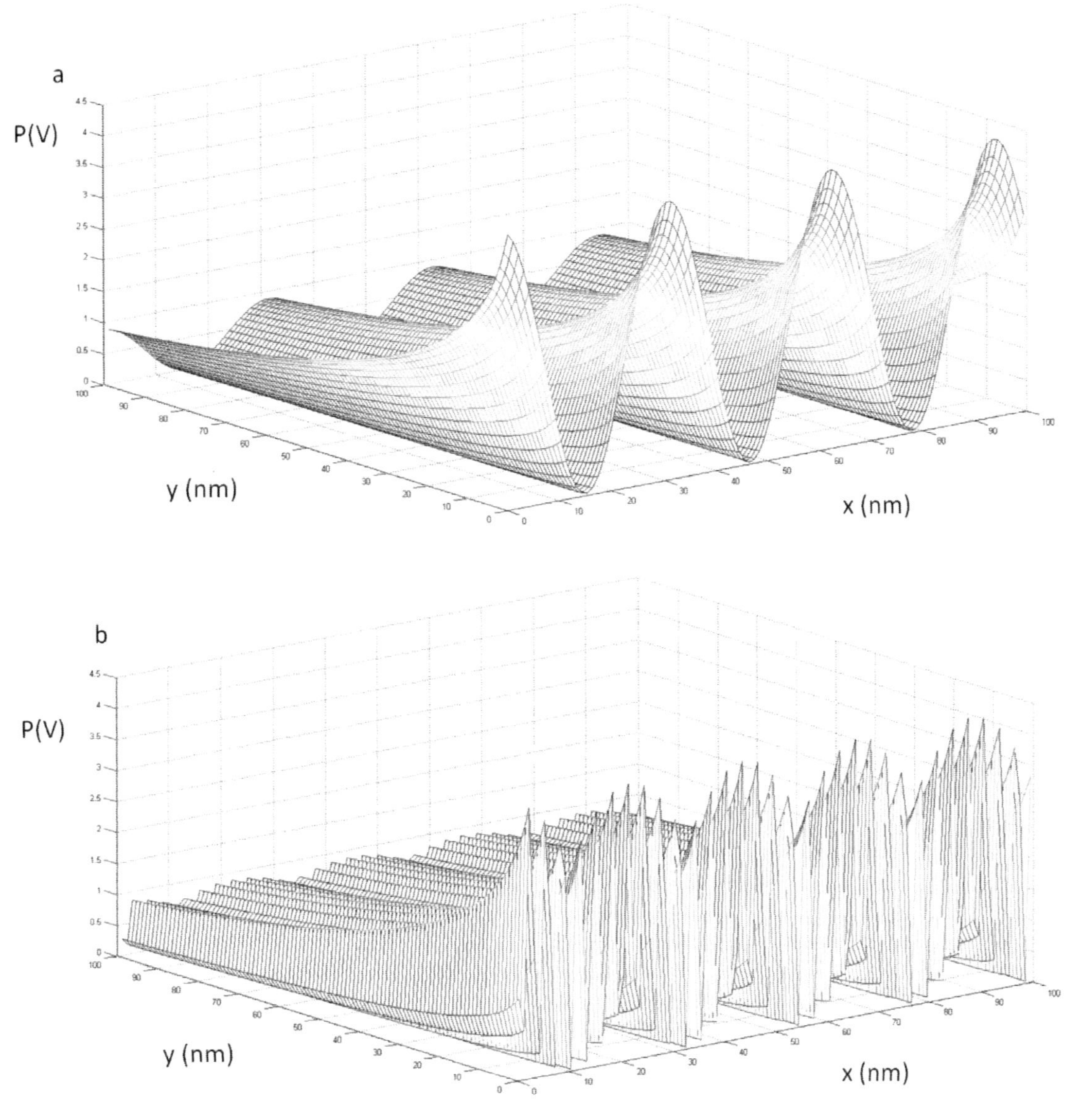


**Figure 3.** Distribution of the electric potential for $\lambda_0 = 10^{-19}$ C, $\frac{1}{4\pi\varepsilon_0} = 10^{10}$ V*m/C. (a), $\mathbf{k}$ = 0.1 nm$^{-1}$. (b), $\mathbf{k}$ = 1 nm$^{-1}$.

## 3. Conclusion

We have presented three methods of calculation of the potential for a periodic structure. The methods were applied for an infinitely long, thin periodically charged thread. They differ qualitatively – in the form of the potential distribution, and quantitatively – in the value of the electric potential. The first method has an unreliable value of the electric potential, but the second one has an unreliable form. The third method has the form of potential distribution from the first method and the value of the electric potential from the second method. Our findings are a step to understanding of how to build a map of the surface electric potential distribution for a periodic structure. It's very important because photonic-plasmonic crystals have periodicity. The method can be usefull for creating materials with necessary design and parameters.

## Acknowledgments
This work was funded by the subsidy allocated to Kazan Federal University for the state assignment in the sphere of scientific activities.

## References
[1] Lopez-Garcia M, Galisteo-Lopez J F, Blanco A, Sanchez-Marcos J, Lopez C and Garcia-Martin A 2010 *Small* **6** (16) 1757-1761
[2] Lopez-Garcia M, Galisteo-Lopez J F, Blanco A, Lopez C and Garcia-Martin A 2010 *Adv. Funct. Mater.* **20** 4338-4343
[3] Galisteo-Lopez J F, Lopez-Garcia M, Lopez C and Garcia-Martin A 2011 *Appl. Phys. Lett.* **99** 083302-1-3
[4] Frederich H, Wen F, Laverdant J, Coolen L, Schwob C and Maitre A 2011 *Optics Express* **19** (24) 24424-24433
[5] Galisteo-Lopez J F, Lopez-Garcia M, Blanco A and Lopez C 2012 *Colloidal Nanoplasmonics* **28** 9174-9179
[6] Frederich H, Wen F, Laverdant J, de Marcillac W D, Schwob C, Coolen L and Maitre A 2014 *Plasmonics* **9** (4) 917-924
[7] Liu T-l, Russel K J, Cui S and Hu E L 2014 *Optics Express* **22** (7) 8219-8225
[8] Joannopoulos J D, Johnson S G, Winn J N and Meade R D 2008 *Photonic crystals: molding the flow of light* (Princeton: Princeton University Press) p 286
[9] Gainutdinov R Kh, Khamadeev M A and Salakhov M Kh 2012 *Phys. Rev.* A **85** 053836
[10] Cole R M, Sugawara Y, Baumberg J J, Mahajan S, Abdelsalam M and Bartlett P N 2006 *Phys. Rew. Lett.* **97** (13) 137401-1-4
[11] Oh Y-J, Park S-G, Kang M-H, Choi J-H, Nam Y and Jeong K-H 2011 *Small* **7** (2) 184-188
[12] Tsvetkov M Y, Khlebtsov B N, Khanadeev V A, Bagratashvili V N, Timashev P S, Samoylovich M I and Khlebtsov N G 2013 *Nanoscale Research Letters* **8** (250) 1-9
[13] Shi L, Yuan X, Zhang Y, Hakala T, Yin S, Han D, Xhu X, Zhang B, Liu X, Torma P, Lu W and Zi J 2014 *Laser and Photonics Reviews* **8** (5) 717-725
[14] Maier S 2007 *Plasmonics: Fundamentals and Applications* (Springer Science+Business Media LLC) p 224
[15] Novotny L and Hecht B 2006 *Principles of nano-optics* (New York: Cambridge Univrsity Press) p 539
[16] Nonnenmacher M, Oboyle M P and Wickramasingle H K 1991 *Appl. Phys. Lett.* **58** 2921-2923
[17] Cohen M, Shavit R and Zalevsky Z 2014 *Scientific Reports* **4** 4096
[18] Kittel C 1996 *Introduction to Solid State Physics* (Wiley) p 673
[19] Koryukin V M, Akhmadeev A A and Salakhov M Kh 2015 *J. Phys.: Conf. Ser.* **613** 012004

Journal of Physics: Conference Series **714** (2016) 012013 doi:10.1088/1742-6596/714/1/012013

# Optical flat bands in 2D waveguide arrays with alternating sign of refraction index

**A I Maimistov[1,2] and I R Gabitov[3,4]**

[1]Department of Solid State Physics and Nanostructures, National Research Nuclear University MEPhI, Moscow, 115409, Russia
[2]Department of Physics and Technology of Nanostructures, Moscow Institute for Physics and Technology, Dolgoprudny, Moscow region, 141700 Russia
[3] Department of Mathematics, University of Arizona, Tucson, Arizona 85721, USA
[4] Skolkovo Institute of Science and Technology, Skolkovo Innovation Center, Moscow, 143026, Russia

E-mail: aimaimistov@gmail.com, gabitov@math.arizona.edu

**Abstract**. We consider the coupled forward and backward waves propagating in two dimensional array of waveguide, which are featured by a positive and negative refraction indexes. The existence of the flat band under certain conditions is demonstrated.

## 1. Introduction

The successes of the last decade in the field of fabrication technology led to the development of optical materials with anticipated spatial characteristics, as well as with properties not found in nature. This includes materials with negative index of refraction. The advent of new materials made it possible to mimic various phenomena of physics by means of optics [1-3]. The most renowned example is transformation optics [4,5], where optical methods allow modelling the phenomenon of optical illusion, invisibility, gravitation and cosmology [3]. Works in this area include those dedicated to the photonic analogue of graphene [6], topological photon states and topological phase transitions [7], the optical Hall effect [8], and other phenomena, the list of which can be considerably extended.

Most optical analogues are based on the following idea: the electrons, localized in individual atoms, are attaining the ability of interatomic tunnelling after consolidation of atoms in the crystal lattice. Thus, they are delocalized, and their energy spectrum is considerably different from the spectrum of free electrons. In the strong-coupling approximation the electrons in the crystal lattice are characterized by Wannier functions that are localized on the lattice nodes and slightly overlap with Wannier functions of the near neighbour nodes. In the case of the electromagnetic field similar effect can be achieved by localizing field in waveguides. In this case, transfer waveguide modes serve as Wannier functions. If waveguides are consolidated in a bundle or in a one/two dimensional array, then due to the distortion of total internal reflection between adjacent waveguides, an electromagnetic wave is delocalized similar to electrons in strongly coupled atomic system. The spectrum of the electromagnetic field is changing; prohibited frequency gaps emerge in the spectrum, which give a basis to address such periodic structures as photonic crystals. However, it should be noted that in general the concept of photonic crystal is not limited to the case of strong coupling. In some cases the electromagnetic wave propagates at an angle to the waveguides axes; this corresponds to the case of weak coupling.

Here we restrict ourselves by the electromagnetic waves propagating along the waveguide only; energy transfer between waveguides takes place only through the field tunnelling between adjacent waveguides (evanescent field). In this situation, the discrete nature of the system of ordered waveguides plays a significant role; this suggests addressing such systems as photonic lattices and introducing the concept of discrete photonics.

In this paper, we consider waveguides only as the "atoms" of the two dimensional photonic lattice, however it should be noted that there are other mechanisms of localization. An example is the localization of radiation in the micro-cavities, acting as an analogue of quantum dots [9,10].

## 2. 2D waveguide arrays

The study of two dimensional electron systems has demonstrated that the presence of the third node in the elementary sell as well as long range interaction in the lattice leads to emerging of a flat sheet (flat band) between conventional zones (see figure 1b). The spectrum of electronic states for two-dimensional square lattice (figure 1a) was obtained in [11,12].

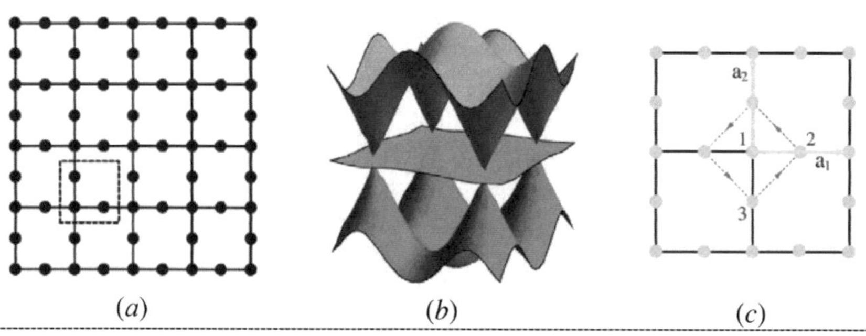


**Figure 1**. Lattice with three nodes in elementary sell (a), surface of the constant energy (see [12]) (b), and Lieb lattice [13] (c).

An analytical expression for the energy is given by

$$\begin{aligned}\varepsilon_1(\mathbf{q}) &= -t\sqrt{4+2\cos(q_x a)+2\cos(q_y a)}\\ \varepsilon_2(\mathbf{q}) &= 0\\ \varepsilon_3(\mathbf{q}) &= t\sqrt{4+2\cos(q_x a)+2\cos(q_y a)}\end{aligned}, \quad (1)$$

here $t$ is integral of hoping, $a$ is lattice parameter (distance between nodes), $q_x$ and $q_y$ are projections of the wave vector on the orthogonal axes. The surface $\varepsilon_2(\mathbf{q})=0$ describes the flat band separating Dirac's points in the corners of the Brillouin zone. A similar structure for energy bands takes place in the case of Lieb lattice [13], figure 1c.

Such optical lattices can be realized by means of waveguides as nodes of the lattice. System of equations describing coupled waves in two dimensional arrays of waveguides shown in figure 2 has the following form

$$\begin{aligned}&i(\partial_t+\sigma_1\partial_z)A_{n,m}+\alpha_1(B_{n,m}+B_{n-1,m})+\alpha_2(C_{n,m}+C_{n,m-1})=0,\\ &i(\partial_t+\sigma_2\partial_z)B_{n,m}+\alpha_1(A_{n,m}+A_{n+1,m})=0,\\ &i(\partial_t+\sigma_3\partial_z)C_{n,m}+\alpha_2(A_{n,m}+A_{n,m+1})=0.\end{aligned} \quad (2)$$

Here $A_{n,m}$ $B_{n,m}$ and $C_{n,m}$ are dimensionless slowly varying amplitudes of coupled waves, sub-indices are numbering elementary sells. Phase matching condition is assumed to be satisfied, $\sigma_j = \pm 1$ stand for the sign of the index of refraction or projection of the wave vector on the axis of the $j$-th waveguide, $\alpha_1$ and $\alpha_2$ are the coupling constants.

Since system of equation (2) is linear, the dispersion relation can be found in a standard way. General expression for dispersion relation in the case of an infinite lattice can be obtained by solving the following cubic equation:

$$(\omega - \sigma_1 k)(\omega - \sigma_2 k)(\omega - \sigma_3 k) - \gamma_1(\omega - \sigma_3 k) - \gamma_2(\omega - \sigma_2 k) = 0, \quad (3)$$

here $\gamma_1 = 4\alpha_1^2 \cos^2(k_x l/2)$, $\gamma_2 = 4\alpha_2^2 \cos^2(k_y l/2)$.

Note that in the limit of an infinite lattice all mode indices turning into continuous variable – two dimensional vector with components $k_x l$, and $k_y l$, here $l$ is a distance between neighboring waveguides.

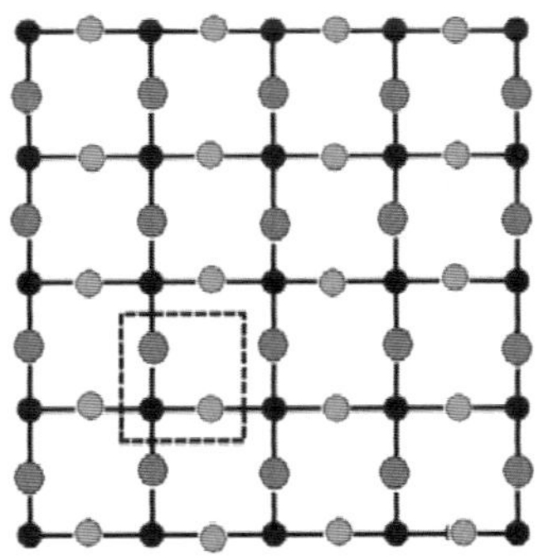

**Figure 2.** 2D array assembled of three types waveguides and ordered in square lattice form.

An analytical solution of equation (3) can be found if one of the waveguides (the first one) has the positive and other two the negative signs of refraction index: $\sigma_1 = 1$, $\sigma_{2,3} = -1$. Corresponding surfaces of the constant frequency (dispersion surfaces) are defined by the following expressions

$$\omega_{1,2}(k, k_x, k_y) = \pm\sqrt{k^2 + 4\alpha_1^2 \cos^2(k_x l/2) + 4\alpha_2^2 \cos^2(k_y l/2)}, \quad (4)$$

$$\omega_3(k, k_x, k_y) = -k. \quad (5)$$

The surface determined by equation (5) at fixed value of $k$ describes a flat band.

For a waveguides with $\sigma_1 = -1$ and $\sigma_{2,3} = +1$, equation (4) remains unchanged, and the surface of the flat band is determined by the equation $\omega_3(k, k_x, k_y) = k$. In the case, when for all waveguides $\sigma_{1,2,3} = +1$, equation (3) has the solution:

$$\omega_{1,2}(k, k_x, k_y) = k \pm 2\sqrt{\alpha_1^2 \cos^2(k_x l/2) + \alpha_2^2 \cos^2(k_y l/2)}, \quad (6)$$

$$\omega_3(k, k_x, k_y) = k. \quad (7)$$

Therefore there is always a flat sheet in the considered lattice of waveguides.

## 3. Conclusion

In this article, we have considered several phenomena of discrete photonic for two-dimensional arrays of waveguides. An analysis of the optical properties of discrete media structured with dielectric micro cavities or dielectric spheroids [9, 10] remained outside of this work.

Note that ~~the~~ above we considered quasi-harmonic electromagnetic waves with carrier frequency $\omega_0$ and with corresponding wave number $k_{z0}$. For this reason, all $\omega$ and $k$ are small with compare to $\omega_0$ and $k_{z0}$ respectively. Hence, expression (2) describes an electromagnetic wave with a narrow spectrum, localized in a small neighborhood of $(\omega_0, k_{z0})$. The existence of the flat band $\omega_3(0, k_x, k_y) = 0$ means that the part of the radiation carried by this mode is not involved in the diffraction or exhibits very small leakage to a neighboring waveguides. In the 2D case, if the frequency belongs to the flat band, then the electromagnetic wave is not experiencing the discrete diffraction.

**Acknowledgments**

We thank our colleagues, N.V. Bykov, Zh.A. Kudyshev and E.V. Kazantseva. The research was supported by Russian Science Foundation (project 14-22-00098).

**References**

[1] Longhi S 2009 *Laser & Photonics Reviews* **3** 243
[2] Yannopapas V 2014 *Int. J. Mod. Phys.* B **28** 1441005
[3] Marini A, Tran Tr X, Roy S, Longhi S and Biancalana F 2014 *Phys. Rev.* A **89** 023840
[4] Shalaev Vl M and Pendry J 2011 *J. Opt.* **13** 020201
[5] Kildishev A V and Shalaev V M 2011 *Physics – Uspekhi* **54** 53–63
[6] Ochiai T and Onoda M 2009 *Phys. Rev.* B **80** 155103
[7] Pendry J B, Schurig D, and Smith D R 2006 *Science* **312** 1780
[8] Onoda M, Murakami Sh and Nagaosa N 2004 *Phys. Rev. Lett.* **93** 083901
[9] Khitrova G, Gibbs H M, Jahnke F, Kira M and Koch S W 1999 *Rev. Mod. Phys.* **71** 1591
[10] Lo'pez-Garci'a M, Galisteo-Lo'pez J F, Lo'pez C and Garci'a-Marti'n A 2012 *Phys. Rev.* B **85** 235145
[11] Yun-tuan Fang, Han-Qing He, Jian-xia Hu, Lin-kun Chen and Zhang Wen 2015 *Phys. Rev.* A **91** 033827
[12] Apaja V, Hyrk¨as M and Manninen M 2010 *Phys. Rev.* B **82** 041402(R)
[13] Weeks C and Franz M 2010 *Phys. Rev.* E **82** 085310

Journal of Physics: Conference Series **714** (2016) 012014 doi:10.1088/1742-6596/714/1/012014

# Femtosecond correlated transient holography in CdS crystal under two-quantum excitation

**V V Samartsev[1,2], A V Leontiev[1] and T G Mitrofanova[1]**

[1]E K Zavoisky Kazan Physical-Technical Institute,Sibirsky Trakt str. 10, Kazan 420029, Russia

[2]Kazan Federal University, Kremlevskaya str. 16, Kazan 420008, Russia

E-mail: samartsev@kfti.knc.ru

**Abstract.** This paper is devoted to the discussion of a possibility of the femtosecond correlated transient holograms recording in CdS crystal under two-quantum excitation.

## 1. Introduction

Let us analyze the situation when the femtosecond correlated transient holograms (FCTH) are recorded into the wave fronts of coherent responses such as the free induction decay (FID) signals formed in CdS crystal under two-quantum excitation. These signals may be used for quantum calculations and are characterized by property of superposition for which there always exists a complete set of eigenstates named as a basis [1]. The property of superposition means that there is no possibility to say which of the two resonant states is realized in practice. For coupled particles there is a correlation (for example, due to the Coulomb interaction of electrons in semiconductor crystal CdS) between them. This correlation leads to the appearance of the correlated states. Electrons in such a state can form an EPR-pair [2].

In our paper [3] we have considered the signals of the correlated free-induction decay (CFID) emitted in CdS crystal under two-quantum excitation by femtosecond 793 nm radiation of a Ti: sapphire laser. The exciting beams propagating in the directions $\mathbf{k}_1$ and $\mathbf{k}_2$ at an angle of 60° to each other induce the appearance of two EPR-gratings of nonequilibrium electric polarization $\mathbf{k}_1 - \mathbf{k}_2$ and $\mathbf{k}_2 - \mathbf{k}_1$ in the crystal. The reason for formation of two gratings is following. The electrons in the CdS crystal need to absorb the energy of two photons of laser radiation in order to pass from valence zone to conduction one. But we should take into consideration only photons belonging to different laser beams. It should be also noted that two quanta of laser energy ($2\hbar\omega_L$) are equal to the sum of quanta ($\hbar\omega_L$) of CFID coherent responses.

## 2. CFID signals in CdS crystal under two-quantum excitation

Since there are two EPR-gratings the process of simultaneous emission of two correlated responses is similar (but only with respect to $\sigma_\pm$ polarizations) to the process of the FID formation in a ruby [4]. Omitting the details of calculation let us write the following formula for the CFID electric field in the case of two-quantum excitation:

$$\mathbf{E}(t,\mathbf{r}) = 2\int_0^{\infty} \mathrm{d}(\Delta\omega) g(\Delta\omega) \sum_j^N W(\Delta\omega,\mathbf{r},t) \mathrm{e}^{-i\mathbf{k}_{\mathrm{CFID}}\mathbf{r}}, \tag{1}$$

where $\Delta\omega = \omega_{\mathrm{e}} - 2\omega_{\mathrm{L}}$ is the detuning, $\omega_{\mathrm{e}}$ is the frequency of the two-quantum energy transition of an electron, $\omega_{\mathrm{L}}$ is the laser frequency, $g(\Delta\omega)$ is the distribution function of the detuning $\Delta\omega$, $W(\Delta\omega,\mathbf{r},t) = -\frac{i}{2}\sin\theta\, \mathrm{e}^{-i\Delta\omega(t-\Delta t)}$, $\Delta t$ is the pulse duration, $\theta$ is the pulse area which under the two-quantum excitation is described as

$$\begin{aligned}\theta &= \hbar^{-1}\Delta t\left[\frac{\langle \mathrm{e}|\mathbf{pE}_0\mathrm{e}^{i(\omega_{\mathrm{L}}t-\mathbf{k}_1\mathbf{r})}|\gamma\rangle\langle\gamma|\mathbf{pE}_0\mathrm{e}^{i(\omega_{\mathrm{L}}t+\mathbf{k}_2\mathbf{r})}|g\rangle}{\hbar(\omega_\gamma-\omega_{\mathrm{L}}+i\Delta_\gamma)} + \frac{\langle \mathrm{e}|\mathbf{pE}_0\mathrm{e}^{i(\omega_{\mathrm{L}}t-\mathbf{k}_2\mathbf{r})}|\gamma\rangle\langle\gamma|\mathbf{pE}_0\mathrm{e}^{i(\omega_{\mathrm{L}}t+\mathbf{k}_1\mathbf{r})}|g\rangle}{\hbar(\omega_\gamma-\omega_{\mathrm{L}}+i\Delta_\gamma)}\right] \\ &= \hbar^{-1}\Delta t Q\mathbf{E}_0^2\left[\mathrm{e}^{-i(\mathbf{k}_1-\mathbf{k}_2)\mathbf{r}} + \mathrm{e}^{-i(\mathbf{k}_2-\mathbf{k}_1)\mathbf{r}}\right].\end{aligned} \tag{2}$$

Here $Q = \frac{\langle \mathrm{e}|\mathbf{p}|\gamma\rangle\langle\gamma|\mathbf{p}|\mathrm{g}\rangle}{\hbar(\omega_\gamma-\omega_{\mathrm{L}}+i\Delta_\gamma)}$ is the parameter of two-quantum excitation, $\langle e|\mathbf{p}|\gamma\rangle$ is the electric dipole moment of the transition, $|\mathrm{g}\rangle$, $|\gamma\rangle$ and $|\mathrm{e}\rangle$ are the ground, intermediate and excited states of an electron, $\omega_\gamma$ and $\Delta_\gamma$ are the frequency and the width of the intermediate state $|\gamma\rangle$, $\mathbf{E}_0$ is the amplitude of the electric field of the exciting pulses, $\mathbf{k}_1$ and $\mathbf{k}_2$ are the wave vectors of the exciting pulses which are at the angle $\varphi$=60° to each other. Then the expression (1) for the electric field of CFID for small $\theta$ takes the form

$$\mathbf{E}(\mathbf{k}_{\mathrm{CFID}},t) = E_0^2\hbar^{-1}\Delta t Q\Phi_{\mathrm{CFID}}\,\mathrm{th}^2\frac{\hbar\omega_e}{2k_{\mathrm{B}}T} f\sum_j^N\left[\mathrm{e}^{-i[\mathbf{k}_{\mathrm{CFID}}-\mathbf{k}_1+\mathbf{k}_2]\mathbf{r}_j} + \mathrm{e}^{-i[\mathbf{k}_{\mathrm{CFID}}-\mathbf{k}_2+\mathbf{k}_1]\mathbf{r}_j}\right], \tag{3}$$

where $\Phi_{\mathrm{CFID}}$ is the form-factor of CFID, which depends on the excitation conditions. The corresponding phase matching conditions follow from formula (3):

$$\begin{aligned}\mathbf{k}_{\mathrm{CFID}} &= \mathbf{k}_1 - \mathbf{k}_2, \\ \mathbf{k}'_{\mathrm{CFID}} &= \mathbf{k}_2 - \mathbf{k}_1.\end{aligned} \tag{4}$$

## 3. Correlated transient holograms

In the experimental work [5] A.K. Rebane with colleagues has shown the possibility of recording of the femtosecond two-quantum holograms in the organic samples. The method was demonstrated on the simplest holographic example – the recording of the frequency-domain gratings, when the first optical beam with the wave vector $\mathbf{k}_1$ and phase $\varphi_1$ is a referent wave and the second beam with the wave vector $\mathbf{k}_2$ and phase $\varphi_2$ is an object one. Since the pulses in both beams are femtosecond we assume that they reach the sample simultaneously. It should be noted that in the case of ordinary transient holography on one-quantum transitions (see review [6]) the exciting beams are divided by time intervals smaller than the $T_1$ and $T_2$ relaxation times. Rebane with colleagues has demonstrated that the time interval between the referent and the object beams can be zero under two-quantum excitation. Let us now examine the possibility of the femtosecond recording and reading of the correlated transient two-quantum holograms in the semiconductor crystal CdS at low temperature. In the area of exciting beams' intersection on the surface of the sample there are the following two optical waves :

*referent wave*

$$\mathbf{E}_1(\mathbf{r}_j,t)=\mathbf{E}_0^{(1)}(\mathbf{r}_j)\exp\{i[\omega_L t-\mathbf{k}_1\mathbf{r}_j-\varphi_1(\mathbf{r})]\}, \tag{5}$$

*object wave*

$$\mathbf{E}_2(\mathbf{r}_j,t)=\sum_{\eta=1}^{m}\mathbf{E}_{0\eta}^{(2)}\exp\{i[\omega_L t-\mathbf{k}_{2\eta}\mathbf{r}_j-\varphi_2(\mathbf{r})]\}. \tag{6}$$

Here $\mathbf{E}_0^{(1)}(\mathbf{r}_j)$, $\omega_L$, $\mathbf{k}_1$ and $\varphi_1(\mathbf{r})$ are the amplitude, laser frequency, wave vector and phase of the referent wave. $\mathbf{E}_{0\eta}^{(2)}$, $\omega_L$, $\mathbf{k}_{2\eta}$ are the amplitude, laser frequency, wave vector of $\eta$-th component in a Fourier transform of the object wave, $\varphi_2(\mathbf{r})$ is the phase of object wave. The electric field of the referent wave together with each plane Fourier-component of the object wave forms the plane Fourier-response of FID. The sum of such Fourier-responses gives us the wave front of the total FID response named as a transient hologram. Now one principal moment should be noted. Since there are two EPR-gratings in CdS crystal then two scattered FID responses will be scattered by these gratings into two opposite directions. So, the formula for CFID electric field can be used for calculation of electric field of the transient holograms. But we will use another representation of the electric field under calculation by formula (2). According [7] the electric field of pulses is conveniently expressed over the wave fronts:

$$\begin{aligned}\mathbf{E}_1(\mathbf{r}_j,t)&=\mathbf{E}_0^{(1)}(\mathbf{r}_j)\exp\{i[\omega_L t-\mathbf{k}_1\mathbf{r}_j-\varphi_1(\mathbf{r})]\},\\ \mathbf{E}_2(\mathbf{r}_j,t)&=\mathbf{E}_0^{(2)}(\mathbf{r}_j)\exp\{i[\omega_L t-\mathbf{k}_2\mathbf{r}_j-\varphi_2(\mathbf{r})]\},\end{aligned} \tag{7}$$

where $\varphi_1(\mathbf{r})$ and $\varphi_2(\mathbf{r})$ are the wave fronts of the first and the second pulses. Here a wave front is a surface of a same phase. Then, the electric field of the correlated transient two-quantum hologram is written as

$$\mathbf{E}_{\mathrm{CTH}}(\mathbf{r},t)=\mathbf{E}_0(\mathbf{r})\left[\mathrm{e}^{-i(\mathbf{k}_1-\mathbf{k}_2-\mathbf{k}_{\mathrm{CTH}})\mathbf{r}-\omega_L(t-\Delta t)-[\varphi_{\mathrm{CTH}}(\mathbf{r})-\varphi_1(\mathbf{r})+\varphi_2(\mathbf{r})]}+\mathrm{e}^{-i(\mathbf{k}_2-\mathbf{k}_1-\mathbf{k}_{\mathrm{CTH}})\mathbf{r}-\omega_L(t-\Delta t)-[\varphi_{\mathrm{CTH}}(\mathbf{r})-\varphi_2(\mathbf{r})-\varphi_1(\mathbf{r})]}\right]. \tag{8}$$

where "CTH" stands for "correlated transient hologram", $E_0(\mathbf{r})=E_0^{(1)}E_0^{(2)}\hbar^{-1}\Delta t Q\Phi_{\mathrm{CTH}}\,\mathrm{th}(\hbar\omega_L/2k_B T)f\,\mathrm{e}^{-(t-\Delta t)/T_2}$, $T_2$ is the total dephasing time (for CdS $T_2$=680 fs), which is due to the exciton-exciton interaction [7]. The corresponding phase matching conditions follow from formula (8):

$$\begin{aligned}\mathbf{k}_{\mathrm{CTH1}}&=\mathbf{k}_1-\mathbf{k}_2,\\ \mathbf{k}_{\mathrm{CTH2}}&=\mathbf{k}_2-\mathbf{k}_1.\end{aligned} \tag{9}$$

The wave fronts $\varphi_{\mathrm{CTH}}(\mathbf{r})$ of the correlated transient holograms are equal to

$$\varphi_{\text{CTH1}}(\mathbf{r}) = \varphi_1(\mathbf{r}) - \varphi_2(\mathbf{r}),$$
$$\varphi_{\text{CTH2}}(\mathbf{r}) = \varphi_2(\mathbf{r}) - \varphi_1(\mathbf{r}). \quad (10)$$

Since the wave front of the referent wave can be zero, the wave fronts of the holograms CTH1 and CTH2 are reversed. The situation is explained in figure1.

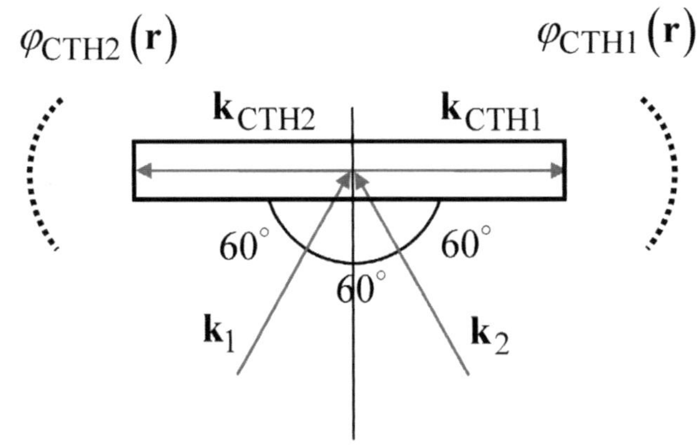


**Figure 1.** Formation of the correlated transient holograms "CTH1" and "CTH2" in CdS crystal.

## 4. Conclusion

Above, the possibility of the correlated transient two-quantum holograms in the CdS crystal is shown. These holograms are reversed. For the spherical wave fronts we obtain [8]

$$\frac{1}{R_{\text{CTH1}}} = \frac{1}{R_1} - \frac{1}{R_2},$$
$$\frac{1}{R_{\text{CTH2}}} = \frac{1}{R_2} - \frac{1}{R_1}, \quad (11)$$

where $R_1$ and $R_2$ are the radius of curvature of the corresponding exciting waves. The holograms "CTH1" and "CTH2" are placed in the wave fronts of signals CFID1 and CFID2. The results of this paper can be used in quantum optics [9].

**Acknowledgments**
This study was supported by RAS Programs "Actual problems of physics of low temperatures" and "Fundamental optical spectroscopy and its applications" and RFBR grans no. 14-02-00041-a and no. 14-02-90000_Bel-a.

**References**

[1] Gorbachev V N and Zhiliba A I 2004 *Physical Foundations of Modern Information Processes* (St.-Petersburg: Institute of Press) (in Russian)
[2] Einstein A, Podolsky B and Rosen N 1935 *Phys. Rev.* **47** 777
[3] Leontiev A V, Lobkov V S, Mitrofanova T G, Samartsev V V and Shmelev A G 2012 *Laser Phys. Lett.* **9** 654
[4] Manykin E A and Samartsev V V 1984 *Optical Echo Spectroscopy* (Moscow: Nauka) (in Russian)
[5] Rebane A, Drobizhev M and Karotki A 2002 *Jour. of Luminescence* **98** 341
[6] Nefediev L A and Samartsev V V 1992 *Laser Phys.* **2** 617
[7] Hillmann F, Voight J and Redlin H 2000 *Appl. Phys. Lett.* **77** 4181
[8] Yevseyev I V, Rubtsova N N and Samartsev V V 2009 *Coherent Transient Processes in Optics* (Moscow: Fizmatlit) (in Russian)
[9] Samartsev V V 2015 *Correlated Photons and Their Applications* (Cambridge, UK: CISP)

# Light bullets

**S V Sazonov**

National Research Centre «Kurchatov Institute», 123182 Moscow, Russia

e-mail: sazonov.sergey@gmail.com

**Abstract.** Light bullets in media with saturating nonlinearity and quadratic nonlinearity under generation of the second terahertz harmonics are considered with the use of the averaged Lagrangian method.

## 1. Introduction

The term "light bullet" for the first time appeared in work [1]. This object denotes a clot of light energy localized in all directions extending in the nonlinear environment with a constant velocity. Focusing nonlinearity, negative dispersion of group velocity (DGV) and diffraction are necessary for its formation. If to distract from some canons of mathematical character, it is possible to call a light bullet a three-dimensional soliton. There are spatial and temporary optical solitons. Spatial solitons are optical bunches, almost unlimited in the direction of propagation and limited in the cross directions. The mechanism of formation of such solitons consists in the fact, that the process of cross self-focusing of a bunch due to nonlinearity of the environment is compensated by its diffraction divergence [2]. Temporary solitons are optical pulses of limited temporary duration. These objects are unlimited in the cross directions and limited in the direction of propagation. Here the nonlinear capsizing of the pulse is compensated by its dispersive spreading. A light bullet can be considered as a symbiosis of spatial and temporary solitons. Compensation of nonlinear self-compression by dispersion (in the direction of propagation) and by diffraction (in the cross directions) leads to the formation of light bullets.

The present work is devoted to a short discussion and physical analysis of the situations at which formation of these optical objects is possible.

## 2. Light bullets in isotropic media

Let the light pulse propagate in an isotropic dielectric in the direction of the $z$-axis. Its electric field $E$ is determined through the complex envelope $\psi$ as follows

$$E(z,\mathbf{r}_\perp,t)=\psi(z,\mathbf{r}_\perp,t)e^{i(\omega t-kz)}+\psi^*(z,\mathbf{r}_\perp,t)e^{-i(\omega t-kz)}, \quad (1)$$

where $\mathbf{r}_\perp$ is the radius vector of transversal coordinates, $t$ is time, $\omega$ and $k$ are the carrier frequency and wave number, respectively.

Using the approach of slowly varying envelopes (SVE) in Maxwell's equations and in the constitutive equations, we come to the wave equation which looks like [3]:

$$i\frac{\partial\psi}{\partial z}+\frac{k_2}{2}\frac{\partial^2\psi}{\partial\tau^2}-F\left(|\psi|^2\right)\psi=\frac{c}{2n\omega}\Delta_\perp\psi\,. \tag{2}$$

Here $k_2=\partial^2k/\partial\omega^2=\partial v_g^{-1}/\partial\omega$ is the parameter of DGV, $\tau=t-z/v_g$ is local time, $c$ is the speed of light in vacuum, $n$ is the refractive index corresponding to the frequency $\omega$, $\Delta_\perp$ is a transversal Laplacian, $F\left(|\psi|^2\right)$ is a positively determined function characterizing the focusing nonlinearity of polarizing response of the medium.

There are various mechanisms of suppression of self-focusing which is necessary for the formation of light bullets. One of them consists in the use of media with saturated nonlinearity [3]. In this case we have

$$F\left(|\psi|^2\right)\sim\frac{|\psi|^2}{1+|\psi|^2/|\psi_s|^2}\approx|\psi|^2\left(1-\frac{|\psi|^2}{|\psi_s|^2}\right), \tag{3}$$

where $|\psi_s|$ is an amplitude which corresponds to the intensity of saturation.

Except saturated nonlinearity, mechanisms connected with gradient wave guides and also with linear dispersion of higher orders, nonlinearity dispersion, Raman nonlinearity, etc. [3] are used.

We won't discuss these mechanisms here, and we will concern only the mechanism connected with saturated nonlinearity with the use of the method of the averaged Lagrangian [4 – 6]. Its essence is as follows. In the beginning the exact one-dimensional solitonic solution of the considered wave equation is found. Then in order to consider the influence of transversal spatial dimensions, the assumption of dependence of some parameters on coordinates in these solutions is introduced. As a result, we have the so-called trial solutions. These solutions are substituted into the Lagrangian, corresponding to the wave equation including derivatives with respect to transversal coordinates. Further, integration of the resulting expression with respect to time is carried out. As a result, we come to an averaged Lagrangian containing the dependence on variable parameters. Writing down the Euler – Lagrange equation for the varied parameters with use of this averaged Lagrangian we will obtain the corresponding differential equations for these parameters.

We will describe equation (3) as follows

$$i\frac{\partial\psi}{\partial z}+\frac{k_2}{2}\frac{\partial^2\psi}{\partial\tau^2}-a|\psi|^2\psi+b|\psi|^4\psi=\frac{c}{2n\omega}\Delta_\perp\psi\,, \tag{4}$$

where $a$ and $b$ are the constants, first of which corresponds to the Kerr nonlinearity, and the second considers its saturation.

At first we will write down the one-dimensional ($\Delta_\perp\psi=0$) solution of equation (4) in the form of a temporary soliton

$$\psi=\frac{1}{\tau_p}\sqrt{\frac{2|k_2|}{a}}\left[\frac{1}{\sqrt{1-q/\tau_p^2}\,\mathrm{ch}\left(2\tau/\tau_p\right)+1}\right]^{1/2}\exp\left(-i\frac{|k_2|z}{2\tau_p^2}\right), \tag{5}$$

where $q=\dfrac{8b|k_2|}{3a^2}$, $\tau_p$ is the temporal pulse duration.

According to (5), for taking into account transversal coordinates we will choose the trial solution of equation (4) in the following form:

$$\psi = \sqrt{\frac{2|k_2|}{a}} Q \left[ \frac{1}{\sqrt{1-qQ^2 \operatorname{ch}(2Q\tau)+1}} \right]^{1/2} \exp(-i\Phi). \tag{6}$$

Here $Q$ and $\Phi$ are functions of coordinates which are subject to search. As in the one-dimensional case, $Q = 1/\tau_p = const.$, and $\Phi \sim z$, now $Q$ and $\Phi$ are the "slow" and "fast" functions of coordinates, respectively.

The Lagrangian that corresponds to equation (10) has an appearance:

$$L = \frac{i}{2}\left( \psi^* \frac{\partial \psi}{\partial z} - \psi \frac{\partial \psi^*}{\partial z} \right) - \frac{a}{2}|\psi|^4 + \frac{b}{3}|\psi|^6 - \frac{k_2}{2}\left|\frac{\partial \psi}{\partial \tau}\right|^2 + \frac{c}{2n\omega}|\nabla_\perp \psi|^2. \tag{7}$$

Substituting (6) into (7), and integrating the resulting expression with respect to $\tau$, we obtain an averaged Lagrangian

$$\Lambda \equiv \frac{a\sqrt{q}}{2|k_2|} \int_{-\infty}^{+\infty} L d\tau = \Lambda_R + \Lambda_D , \tag{8}$$

where the refraction $\Lambda_R$ and diffraction $\Lambda_D$ parts of an averaged Lagrangian have respectively an appearance

$$\Lambda_R = \rho \frac{\partial \varphi}{\partial z} + \rho \frac{(\nabla_\perp \varphi)^2}{2} - \frac{c|k_2|}{2n\omega q}(\rho - \operatorname{th}\rho), \tag{9}$$

$$\Lambda_D = \left(\frac{c}{n\omega}\right)^2 G(\rho)(\nabla_\perp \rho)^2 , \tag{10}$$

$$\rho = \frac{1}{2}\ln\left(\frac{1+\sqrt{q}Q}{1-\sqrt{q}Q}\right), \qquad \varphi = \frac{c}{n\omega}\Phi, \tag{11}$$

$$G(\rho) = \operatorname{cth}\rho + \frac{2}{\operatorname{sh}\rho \operatorname{ch}\rho} - \frac{\rho}{\operatorname{sh}^2\rho} + \frac{\pi^2 + 3\rho^2}{3\operatorname{sh}^3\rho \operatorname{ch}\rho} - \rho \frac{\pi^2 + \rho^2}{3\operatorname{sh}^4\rho \operatorname{ch}^2\rho}. \tag{12}$$

The refraction part of the Lagrangian corresponds to the approach of geometrical optics for solitons [4, 6]. Writing down the Euler – Lagrange equations for $\rho$ and $\varphi$ with the use of (8) – (12), we will have

$$\frac{\partial \rho}{\partial z} + \nabla_\perp (\rho \nabla_\perp \varphi) = 0, \tag{13}$$

$$\frac{\partial \varphi}{\partial z} + \frac{(\nabla_\perp \varphi)^2}{2} + f(\rho) = \left(\frac{c}{n\omega}\right)^2 D_s, \tag{14}$$

where

$$D_s = 2(\Delta_\perp \rho) G(\rho) + (\nabla_\perp \rho)^2 G'(\rho), \qquad f(\rho) = -\frac{c|k_2|}{2n\omega q}\operatorname{th}^2\rho. \tag{15}$$

When the right-hand part in (14) equals zero, system (13), (14) is formally similar to the system of equations of hydrodynamics for two-dimensional ideal liquid. Thus, (13) represents the continuity equation, and (14) is the Cauchy's integral. The right-hand part of (14) describes the influence of diffraction on the propagation of a pulse [7]. There are bases to believe that diffraction is capable to compensate the self-focusing of a pulse.

Equation (13) is satisfied by the solutions corresponding to an axial-symmetric light bullet:

$$\rho = \rho_0 \frac{R_0^2}{R^2(z)} F\left(r / R(z)\right), \qquad \varphi = f_1(z) + \frac{r^2}{2} \frac{R'(z)}{R(z)}. \tag{16}$$

Here $r$ is a radial component of a cylindrical coordinate system, making meaning of a distance from the focal $z$-axis to a supervision point, $R$ is the characteristic transversal radius (aperture) of a light bunch depending on $z$, $R_0$ and $\rho_0$ are the radius and $\rho$ of the bunch at the input into the medium (at $z = 0$), $F$ is the function defined after substitution (16) into (14).

Following Ref. [2], we will consider a transversal profile of a soliton as a Gaussian, having chosen $F$ in the form:

$$F = \exp\left(-r^2 / R^2\right). \tag{17}$$

Besides, by the substitution (16) and (17) into (14) we will take into account following approximation [2]

$$\mu = r^2 / R^2 << 1. \tag{18}$$

This approximation enables a series expansion up to the first degree of the small parameter $\mu$. Then, equating expressions in the left-hand and right-hand parts of equation (16) respectively at the zero point and by the first degrees of $\mu$ leads to the following differential equations

$$f_1'(z) = \frac{c|k_2|}{2n\omega q} \mathrm{th}^2\left(\rho_0 \frac{R_0^2}{R^2}\right) + F_D(R), \tag{19}$$

$$R''(z) = -\frac{\partial U}{\partial R}, \tag{20}$$

$$U(R) = U_R(R) + U_D(R), \tag{21}$$

$$U_R = -\frac{c|k_2|}{2n\omega q} \mathrm{th}^2\left(\rho_0 \frac{R_0^2}{R^2}\right). \tag{22}$$

The functions $F_D(R)$ and $U_D(R)$ have very bulky appearance. Therefore, we don't write out them here. The function $U_D(R)$ is positive and monotonously decreases by the increase of the variable $R$.

Equation (20) is similar to Newton's second law for a particle of unit mass with the coordinate $R$ where $z$ plays the role of time. To this movement corresponds the "potential energy" $U$.

The dependence $U(R)$ is shown in figure 1. The local minimum of the function $U(R)$ corresponds to an equilibrium aperture $R_m$ of light bullets. It is easy to see that under the conditions $R'(0) = 0$ and $R_0 > R_c$ a light bullet is stable. As can be seen from the second of expressions (16), the value $R'(z)$ defines the curvature of wave fronts of the pulse. Therefore, wave fronts of a light bullet at $R = R_m$ are flat.

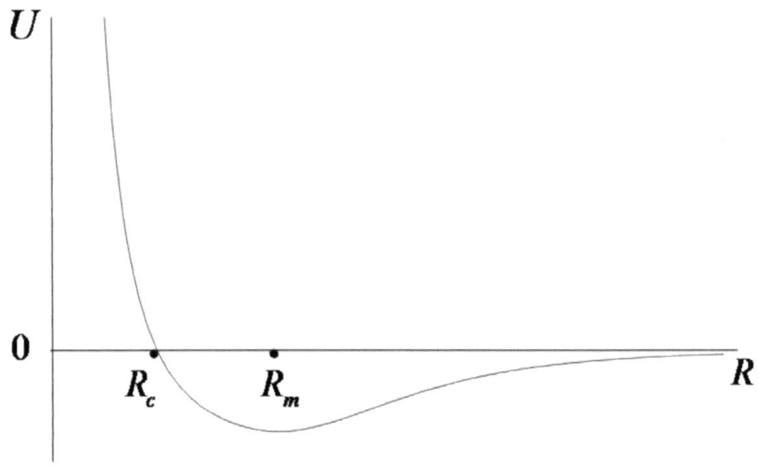


**Figure 1.** Dependence $U(R)$.

### 3. Light bullets in uni-axial crystals at second harmonic generation

Under the fulfillment of the conditions of phase synchronism and group synchronism, the set of equations for second harmonic generation looks like [8]

$$i\frac{\partial\psi_1}{\partial z}+\frac{k_2^{(1)}}{2}\frac{\partial^2\psi_1}{\partial\tau^2}-d_1\psi_1^*\psi_2=\frac{c}{2n\omega}\Delta_\perp\psi_1, \quad (23)$$

$$i\frac{\partial\psi_2}{\partial z}+\frac{k_2^{(2)}}{2}\frac{\partial^2\psi_2}{\partial\tau^2}-d_2\psi_1^2=\frac{c}{4n\omega}\Delta_\perp\psi_2. \quad (24)$$

Here $d_1$ and $d_2$ are the coefficients of quadratic nonlinearity proportional to the nonlinear susceptibility of the second order $\chi^{(2)}$ for both harmonics, $k_2^{(1)}$ and $k_2^{(2)}$ are the DGV coefficients of the first and second harmonics respectively.

Let the condition

$$k_2^{(2)}=2k_2^{(1)} \quad (25)$$

be satisfied. Then set (23), (24) possesses one-dimensional ($\Delta_\perp=0$) solitonic solutions [9]:

$$\psi_1=\pm\frac{3k_2}{4\tau_p^2}\sqrt{\frac{2}{d_1d_2}}\exp\left(i\frac{k_2z}{2\tau_p^2}\right)\mathrm{sech}^2\left(\frac{\tau}{2\tau_p}\right), \quad (26)$$

$$\psi_2=-\frac{3k_2}{4\tau_p^2d_1}\exp\left(i\frac{k_2z}{\tau_p^2}\right)\mathrm{sech}^2\left(\frac{\tau}{2\tau_p}\right). \quad (27)$$

Here $k_2\equiv k_2^{(1)}$.

Now we will consider the transversal dynamics determined by the right-hand parts in (23) and (24). After the scale transformation $\psi_1=\sqrt{d_1/(2d_2)}\Phi_1$, $\psi_2=\Phi_2$, we will write down the Lagrangian which corresponds to set (23), (24):

$$L = L_1 + L_2 + L_{\text{int}}, \tag{28}$$

where

$$L_1 = \frac{i}{2}\left(\Phi_1^* \frac{\partial \Phi_1}{\partial z} - \Phi_1 \frac{\partial \Phi_1^*}{\partial z}\right) - \frac{k_2}{2}\left|\frac{\partial \Phi_1}{\partial \tau}\right|^2 + \frac{c}{2n\omega}\left|\nabla_\perp \Phi_1\right|^2, \tag{29}$$

$$L_2 = \frac{i}{2}\left(\Phi_2^* \frac{\partial \Phi_2}{\partial z} - \Phi_2 \frac{\partial \Phi_2^*}{\partial z}\right) - k_2\left|\frac{\partial \Phi_2}{\partial \tau}\right|^2 + \frac{c}{4n\omega}\left|\nabla_\perp \Phi_2\right|^2, \tag{30}$$

$$L_{\text{int}} = -\frac{d_1}{2}\left(\Phi_1^{*2}\Phi_2 + \Phi_1^2\Phi_2^*\right). \tag{31}$$

According to (26) and (27), we will choose trial solutions in the form

$$\Phi_1 = \pm\frac{6k_2}{d_1}\rho^{2/3}\exp\left(-i\frac{n\omega}{c}\varphi\right)\text{sech}^2\left(\rho^{1/3}\tau\right), \tag{32}$$

$$\Phi_2 = -\frac{3k_2}{d_1}\rho^{2/3}\exp\left(-2i\frac{n\omega}{c}\varphi\right)\text{sech}^2\left(\rho^{1/3}\tau\right). \tag{33}$$

Here $\rho$ and $\varphi$ are unknown functions of coordinates.

After substitution (32) and (33) into (28) – (31) and integration with respect to $\tau$ we will have

$$\int_{-\infty}^{+\infty} L dt = 216\left(\frac{k_2}{\gamma}\right)^2 \Lambda,$$

where the averaged Lagrangian is

$$\Lambda = \frac{\rho}{3}\left[\frac{\partial \varphi}{\partial z} + \frac{\left(\nabla_\perp \varphi\right)^2}{2}\right] + \frac{2ck_2}{5n\omega}\rho^{5/3} + \left(1 + \frac{\pi^2}{30}\right)\left(\frac{c}{n\omega}\right)^2 \frac{\left(\nabla_\perp \rho\right)^2}{36\rho}. \tag{34}$$

Writing down the corresponding Euler – Lagrange equations for $\rho$ and $\varphi$, again we will come to system of hydrodynamic type (13,14), where

$$D_s = \frac{1}{3}\left(1 + \frac{\pi^2}{30}\right)\frac{\Delta_\perp \sqrt{\rho}}{\sqrt{\rho}}, \qquad f(\rho) = \frac{2c}{n\omega}k_2\rho^{2/3}. \tag{35}$$

By analogy with the previous section, we will put $F = \exp\left(-3r^2/2R^2\right)$. Using expressions (16) and approximation (26) we will come to equation (20) for $R$, where

$$U = \frac{3ck_2}{n\omega}\rho_0^{2/3}\frac{R_0^{4/3}}{R^{4/3}} + \frac{9}{4}\left(1 + \frac{\pi^2}{30}\right)\left(\frac{c}{n\omega}\right)^2 \frac{1}{R^2}. \tag{36}$$

From this it follows that $U(R)$ possesses a minimum, only if $k_2 < 0$. Here the dependence of $U(R)$ qualitatively coincides with that represented in figure 1. The position of the minimum of $U$ there corresponds to the value of the "bullet" radius:

$$R_m = 0.295\left(\frac{c}{n\omega|k_2|}\right)^{3/2}\frac{\tau_0^3}{R_0^2}.$$

Here $\tau_0 = 1/\rho_0$.

Thus, by the propagation of a bullet its transversal size oscillates near the value $R_m$. Putting in the last expression $R_m = R_0$, we obtain

$$R_0 = 0.706\sqrt{\frac{c}{n\omega|k_2|}}\tau_0. \tag{37}$$

In this case oscillations of the bullet parameters don't happen, and the transversal size of a bullet is proportional to its temporary duration.

For the results received in this section to be correct, the conditions of phase synchronism and group synchronism are important. Besides, equality (25) has to be carried out. Most likely, it is impossible to strictly meet all these conditions simultaneously. However, it can be made approximately, if frequencies $\omega$ and $2\omega$ lie below characteristic frequencies of the resounding absorption. This is the case, for example, for the terahertz range. In this case the wave number $k \approx n_0\omega/c + s\omega^3$, where $s$ is a constant, and the second term is little in comparison with the first one. Then, approximately, it is possible to consider phase and group velocities of both harmonics by equal $c/n_0$. At the same time, $k_2 \equiv \partial^2 k/\partial\omega^2 = 6s\omega$. Then condition (25) is satisfied automatically. However, in the presence of only temporary dispersion the parameter $s$ is positive. Therefore, $k_2 > 0$. At the same time, it is necessary for the formation of "bullets", that $k_2 < 0$. In a micro-dispersive (granulated) medium spatial dispersion is important. Then under certain conditions the parameter $s$ can become negative [10]. Thereof, we have $k_2 < 0$ at simultaneous performance of condition (25). Then the formation of bullets on both harmonics is possible.

## 4. Concluding remarks

Let us note, that light bullets in quadratic-nonlinear media were predicted earlier, in particular with the use of variation approach (see [3] and the references quoted there). However, the reduction of such a research to the equations of hydrodynamic type (13), (14) and the further search for their solutions is executed here for the first time.

Unlike other approaches the method of an averaged Lagrangian allows separating the nonlinear refraction determined by the type of the function in (14) [6] from the diffraction determined by the type of the right-hand part in (14) [7].

Thus, the approach offered above, describes dynamics of spatial-temporal solitons with the possibility of light bullets formation more exhaustively.

## Acknowledgments

This work is supported by the Russian Foundation for Basic Research (grant No. 16-02-00453-a).

## References

[1] Silberberg Ya 1990 *Optics Letters* **15** 1282
[2] Akhmanov S A, Sukhorukov A P and Khokhlov R V 1968 *Sov. Phys.-Usp.* **10** 609
[3] Kivshar Yu S and Agrawal G P 2003 *Optical solitons* (New York: Academic Press)
[4] Zhdanov S K and Trubnikov B A 1987 *Sov. Phys. JETP* **66** 904

[5] Anderson D, Desaix M, Lisak M and Quiroga-Teixeiro M L 1992 *J. Opt. Soc. Am.* B **9** 1358
[6] Zhdanov S K and Trubnikov B A 1991 *Quasi-Gaseous Unstable Media* (in Russian, Moscow: Nauka)
[7] Sazonov S V 2004 *JETP* **98** 1237
[8] Karamzin Yu N, Sukhorukov A P and Trofimov V A 1989 *Mathematical Modeling in Nonlinear Optics* (in Russian, Moscow: Lomonosov Moscow State University)
[9] Karamzin Yu N and Sukhorukov A P 1974 *Sov. JETP Lett.* **20** 339
[10] Sazonov S V 1995 *Opt. Spectrosc.* **79** 260

# Magnetic control of waveguide modes of Bragg structures

**D A Sylgacheva[1,2], N E Khokhlov[1,2], A N Kalish[1,2] and V I Belotelov[1,2]**

[1]Lomonosov Moscow State University, Moscow, 119991, Russia
[2]Russian Quantum Center, Skolkovo, Moscow Region, 143025, Russia

E-mail: sylgacheva.darjja@physics.msu.ru

**Abstract**. We present the study of the waveguide modes of one-dimensional magnetic photonic crystals with in-plane-magnetized layers. There is a magneto-optical effect of non-reciprocity for the TM-modes propagating along the layers perpendicularly to the magnetization. Due to the non-reciprocity the phase velocity of the modes changes with magnetization reversal. Comparison of the effect in the non-magnetic photonic crystal with additional magnetic layer on top and a magnetophotonic crystal with altering magnetic layers shows that the effect is greater in the first case due to the higher asymmetry of the claddings of the magnetic layer. This effect is important for the light modulation with external magnetic field in waveguide structures and may be used for design of novel types of the magneto-optical devices, sensors of magnetic field or biosensors.

## 1. Introduction

Modern requirements for optical information rates stimulate researchers to discover new methods of light control on submicron scales. Multilayer planar waveguides open new perspectives in this area. Systems of planar waveguides are used in fiber-optic communication WDM system, for example, and other practical applications [1-7]. The combinations of planar waveguides and Bragg structures or 1D photonic crystals (PCs) allow one to obtain the desired transmission spectrum or to specify the necessary eigenmodes dispersion characteristics of the whole structure [5-6]. Also, such structures broaden the applications of waveguides, including optical switchers, waveguide Mach-Zehnder interferometer, etc. [7].

Incorporation of the magnetic media layers into PC represents special interest. In this case, the magnetization of layers modifies the light's polarization and/or phase velocity [8-13] and the resonant enhancement of magneto-optical effects may be observed [15-18]. The authors of Refs. [19-20] investigated the localized states, the so-called optical Tamm states, at the interface between PCs containing magnetic layers and magneto-optical effect enhancement at normal light incidence. However, the case of the light waveguided along the PC's layers is also of great interest. The dispersion of the guided modes in single transversely-magnetized single film depends linearly on the magnetization [21-25]. A similar effect in transversely-magnetized PC's layers is expected. But there may be some special features since the PC's modes are coupled to each other. The topic has not been studied yet to the best of our knowledge.

Here we theoretically investigate the influence of transverse magnetic field on the waveguide TM-modes of a PC containing magnetic layers. Two types of PCs with an additional top layer are considered (figure 1). The structure of type I includes the magnetic layer on top only (figure 1(b)), while the structure of type II is represented as a PC with magnetized layers within its period (figure

1(c)). Here we extend the theory for waveguide modes' dispersion in the case of transverse magnetization of layered periodic structures. The solution of the dispersion equation shows that the transverse magnetization gives rise to magneto-optical non-reciprocity.

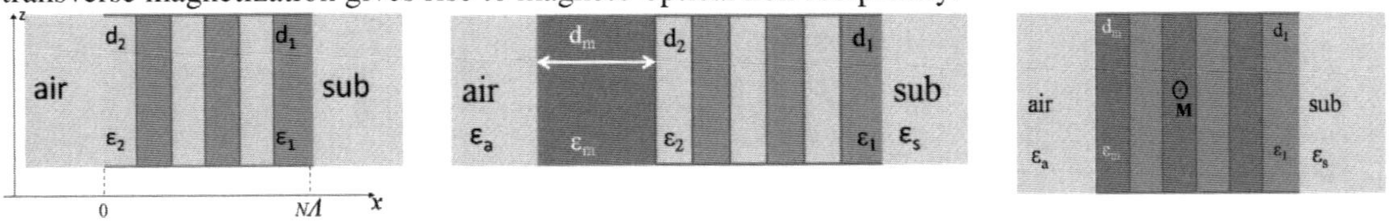


(a) (b) (c)

**Figure 1.** Schemes of 1D photonic crystals: (a) non-magnetic PC sandwiched between semi-infinite air and semi-infinite substrate ("sub"); (b) non-magnetic PC with magnetic layer on top – investigated structure of type I; (c) magnetic PC with altering magnetic/non-magnetic layers – investigated structure of type II; $\varepsilon_1$, $\varepsilon_2$, $\varepsilon_m$ are permittivities of layers with thickness $d_1$, $d_2$, $d_m$ respectively; $d_1$, $d_2$ are quarter wavelength thicknesses, and so is $d_m$ for type II. Magnetic field $\boldsymbol{B}$ is along $y$-axis.

## 2. Method of dispersion calculation

The properties of the waveguide modes in the layers of non-magnetic 1D PC were theoretically studied in [26-32]. According to the developed approach the dispersion law for PC's waveguide modes can be found by constructing a scattering matrix equation and finding the condition for its non-trivial solutions.

We briefly outline the main points of this approach below. Let's consider a PC structure sandwiched between various semi-infinite media (figure 1(a)). The waves propagating forward and backward in the air (with complex amplitudes $f_l$ and $b_l$) and in the substrate ($f_r$ and $b_r$) are related to each other through the linear equation with transfer matrix $\hat{T}$:

$$\begin{pmatrix} f_r \\ b_r \end{pmatrix} = \hat{T} \begin{pmatrix} f_l \\ b_l \end{pmatrix}. \tag{1}$$

The expression for amplitudes of the waves outgoing from the PC through waves incident on the PC has the form with the scattering matrix $\hat{S}$:

$$\begin{pmatrix} f_r \\ b_l \end{pmatrix} = \hat{S} \begin{pmatrix} f_l \\ b_r \end{pmatrix}. \tag{2}$$

The elements of $\hat{S}$ can be expressed through the elements of $\hat{T}$, $t_{ij}$, as follows:

$$\hat{S} = \begin{pmatrix} 1/t_{11} & t_{21}/t_{11} \\ -t_{12}/t_{11} & (t_{11}t_{22} - t_{12}t_{21})/t_{11} \end{pmatrix}^{-1}. \tag{3}$$

The $\hat{S}$-matrix is a function of frequency and quasiwavenumber.

To find eigenmodes we assume that waves incident on the PC are absent, i.e. $f_l=b_r=0$. According to equation (2), the dispersion equation is obtained by equating to zero the determinant of the inverse scattering matrix:

$$\det(\hat{S}^{-1})=0. \tag{4}$$

According to equation (3), $\det(\hat{S}^{-1})=t_{22}/t_{11}$. Consequently,

$$t_{22}=0. \tag{5}$$

To take into account the effect of the magnetization of the layers we represent the medium dielectric tensor in a coordinate system where the $x$-axis is orthogonal to the layers and modes propagate along the $z$-axis:

$$\hat{\varepsilon}_M = \begin{pmatrix} \varepsilon_x & ig_z & -ig_y \\ -ig_z & \varepsilon_y & ig_x \\ ig_y & -ig_x & \varepsilon_z \end{pmatrix}, \tag{6}$$

where $\boldsymbol{g}$ is the gyration vector, linear in the magnetization $\boldsymbol{M}$ and co-directional with it. Here we assume the mediums are isotropic, $\varepsilon_x=\varepsilon_y=\varepsilon_z=\varepsilon$, and $\boldsymbol{M}$ is in the transverse configuration, i.e. $g_x=g_z=0$, $g_y=g$.

The PC is formed by two types of layers, "1" and "2", so:

$$\varepsilon(x) = \begin{cases} \varepsilon_2, & 0 < x < b \\ \varepsilon_1, & b < x < \Lambda \end{cases},$$

$$g(x) = \begin{cases} g_2, & 0 < x < b \\ g_1, & b < x < \Lambda \end{cases}.$$

where $\Lambda$ is the period of the structure; $\varepsilon(x)=\varepsilon(x+\Lambda)$ and $g(x)=g(x+\Lambda)$. The complex amplitude of the magnetic field inside each homogeneous layer $\alpha$ ($\alpha$=1,2) of $n$-th PC unit cell can be represented as the sum of forward and backward plane waves:

$$H_y(x,z) = \left[a_n^{(\alpha)} e^{ik_{\alpha x}(x-n\Lambda)} + b_n^{(\alpha)} e^{-ik_{\alpha x}(x-n\Lambda)}\right] e^{i\beta z},$$

where $k_{\alpha x} = \sqrt{k_0^2\,\varepsilon_\alpha - \beta^2}$, $\beta$ is the mode wavenumber, and $k_0$ is the vacuum wavenumber.

The continuity conditions for components $E_z$ and $H_y$ for $x=(n-1)\Lambda$ and $x=(n-1)\Lambda+b$ leads to the matrix equation:

$$\begin{pmatrix} a_{n-1}^{(1)} \\ b_{n-1}^{(1)} \end{pmatrix} = \begin{pmatrix} A & B \\ C & D \end{pmatrix} \begin{pmatrix} a_n^{(1)} \\ b_n^{(1)} \end{pmatrix}. \tag{7}$$

In this equation, the elements within the linear approximation in $\boldsymbol{g}$ are represented as follows:

$$A = e^{-ik_{1x}a}\left(\cos(k_{2x}b) - i\frac{1}{2}\sin(k_{2x}b)\left[\frac{k_{1x}\varepsilon_2}{k_{2x}\varepsilon_1} + \frac{k_{2x}\varepsilon_1}{k_{1x}\varepsilon_2}\right]\right),$$

$$B = e^{ik_{1x}a}\left(i\frac{1}{2}\sin(k_{2x}b)\left[\frac{k_{1x}\varepsilon_2}{k_{2x}\varepsilon_1} - \frac{k_{2x}\varepsilon_1}{k_{1x}\varepsilon_2} + \frac{2i\beta\varepsilon_2}{k_{2x}}\left(\frac{g_2}{\varepsilon_2^2} - \frac{g_1}{\varepsilon_1^2}\right)\right]\right),$$

$$C = e^{-ik_{1x}a}\left(-i\frac{1}{2}\sin(k_{2x}b)\left[\frac{k_{1x}\varepsilon_2}{k_{2x}\varepsilon_1} - \frac{k_{2x}\varepsilon_1}{k_{1x}\varepsilon_2} - \frac{2i\beta\varepsilon_2}{k_{2x}}\left(\frac{g_2}{\varepsilon_2^2} - \frac{g_1}{\varepsilon_1^2}\right)\right]\right), \tag{8}$$

$$D = e^{ik_{1x}a}\left(\cos(k_{2x}b) + i\frac{1}{2}\sin(k_{2x}b)\left(\frac{k_{1x}\varepsilon_2}{k_{2x}\varepsilon_1} + \frac{k_{2x}\varepsilon_1}{k_{1x}\varepsilon_2}\right)\right),$$

where $a = \Lambda\text{-}b$. Equation (6) shows the transfer matrix for one PC unit cell. So, the transfer matrix for $N$ unit cells can be found as:

$$\hat{T} = \begin{pmatrix} A & B \\ C & D \end{pmatrix}^N \tag{9}$$

The number N of periods of the PC affects the number of guided modes [26-28]. Equations (5), (8) and (9) provide the transcendental dispersion equation which with permittivity tensor (6) gives the opportunity to find magneto-optical features the different PC modes.

The approach is easily generalized to the cases of structures of types I and II, by applying the standard T-matrix formalism for construction of the total transfer matrix for the whole structure.

The modes can be excited by means of a diffraction grating deposited on the top of the structure. At this the dispersion curves can be translated to Brillouin zones associated with the diffraction grating, so the modes that appear inside the light cone are excited directly by the incident light. They form the resonant features of the optical response of the structure, and of the magneto-optical response as well. Figure 2 shows the calculated dispersion diagram for the structure covered with metallic grating and the transverse magneto-optical Kerr effect (TMOKE). The latter is a magnetization-induced change in reflectance. The TMOKE was calculated numerically by the Rigorous Coupled Wave Analysis (RCWA) [33]. It exhibits resonant features associated with the excitation of modes. Comparison of figures 2(a) and 2(b) reveals good agreement between calculations by equation (5) and numerical calculations by RCWA.

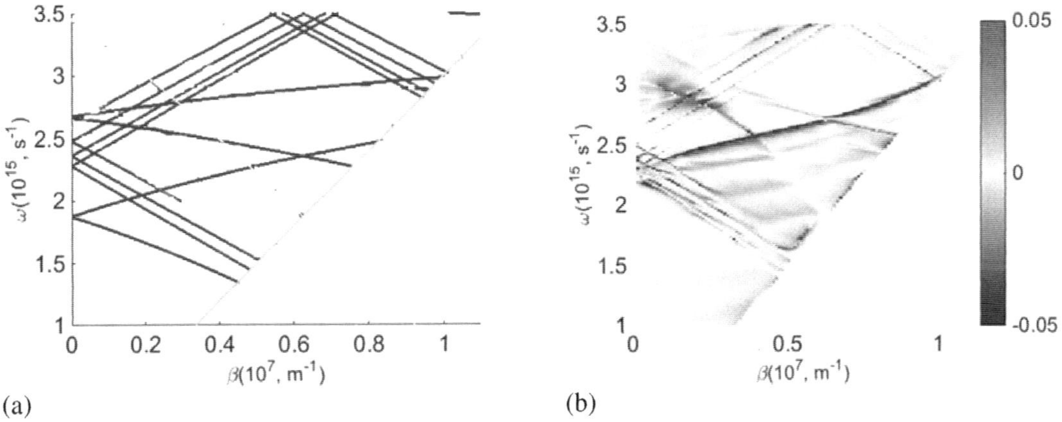


**Figure 2.** (a) Dispersion curves obtained from equation (5) (only the area inside the light cone is shown); (b) the value of the transverse Kerr effect in reflection calculated numerically by RCWA. Structure of type I covered with thin metallic grating is considered.

## 3. The magneto-optical properties of the waveguide modes

The solutions of (5) for structures of types I and II (figure 1 (b,c)) give non-zero effect of the magneto-optical non-reciprocity. It arises due to non-equal wave's phase velocities for opposite magnetization directions and can be characterized by the value:

$$\delta = \frac{\beta(+\mathbf{g}) - \beta(-\mathbf{g})}{\beta(+\mathbf{g}) + \beta(-\mathbf{g})}. \tag{10}$$

The non-zero value $\delta$ means the non-zero magneto-optical effects for the mode. And the problem for optimal structure configuration for higher $\delta$ arises. We have calculated $\delta$ for PCs of types I and II with number of periods $N$=7 (figure 3).

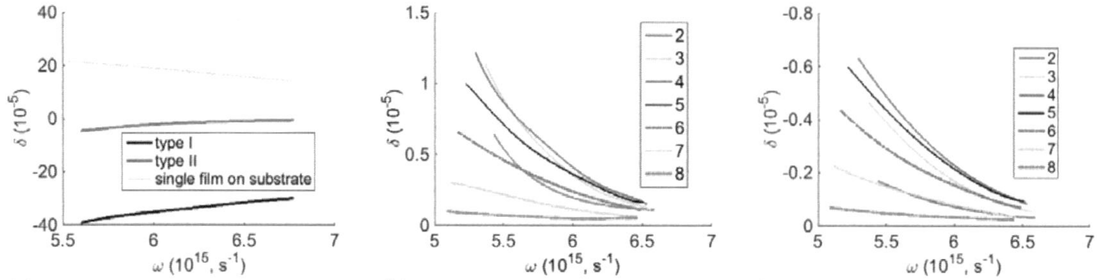


(a) (b) (c)

**Figure 3.** (a) Plots of magneto-optical non-reciprocity value $\delta$ for waveguide mode "1", localized mainly inside the top layer for structures of type I (bottom line), type II (middle line) and single magnetic film on substrate (top line); (b, c) the same value $\delta$ for structures of type I and type II respectively for other modes from the same mode family.

The magnitude of the effect for structure of type I is greater than for the single film (figure 3(a)) due to the greater wave localization inside the magnetic top layer. A more interesting fact is the greater effect in structure of type I than in type II for other modes localized inside the PC's non-top layers. It seems unexpected due to PC of type I is non-magnetic in opposite to magnetic PC of type II. The feature originates due to higher asymmetry of cladding of magnetic in type I rather than in type II. Detailed consideration of transverse magneto-optical Kerr effect and the role of asymmetry of claddings of single magnetic film is presented in [34]. Experimental studies of the transverse magneto-optical Kerr effect for waveguide modes of PC's with a microresonator layer have been performed recently and the resonant effect enhancement was demonstrated for the modes excitations [35]. Another consequence of the asymmetry is the difference in the signs of $\delta$ in the magneto-optical modes "2-8" in the structures of the type I and II (Figure 3 (b,c)). The situation is similar to that considered in [23].

The magnitude of $\delta$ decreases at higher frequencies. This can be explained by the increase in the density of modes in this spectral region. The interaction between the modes decreases with the frequency growth. It leads to the convergence of the eigen-frequencies of waveguide modes. Therefore, the configuration of the mode field distribution becomes closer to the configuration of the field of the single symmetric waveguide, for which the value of $\delta$ is low.

The observed magneto-optical non-reciprocity effect for modes can be revealed in the far-field via the resonant TMOKE that is observed when a structure is covered with diffraction grating, as discussed above (see figure 2). It is the nonreciprocity effect that is the nature of the resonant TMOKE in grating structures [23]. The influence of the mode features on the far-field magneto-optical response provides the opportunity of applying the considered features, i.e., for purposes of magnetic field sensing. The modes are influenced differently by the magnetization of different layers, so the volume sensing of magnetic field becomes possible.

## 4. Conclusions

We extended the theory for the dispersion of waveguide modes in layered dielectric structures with transversely-magnetized layers. It is shown that the dispersion equation contains additional terms proportional to the material gyration. The solutions show the non-zero magneto-optical non-reciprocity. The effect is 2-2.5 times higher for the non-magnetic PC with an additional magnetic top layer as compared to PC with magnetic layers. The effect on the guided mode localized inside the top magnetic layer of a non-magnetic PC is two times greater than for a single magnetic film. Thus, the magneto-optical properties of waveguide modes are largely determined by the claddings, their arrangement and configuration. The considered effect of non-reciprocity is observed in experiment as resonances of transverse magneto-optical Kerr effect. The considered magneto-optical effect's features may be used for design of different magneto-optical devices, sensors of magnetic field or biosensors.

**Acknowledgments**
The work is supported by the Russian Science Foundation (project No. 14-32-00010).

**References**

[1] Freeman R 2002 *Fiber Optic Systems for Telecommunications* (Hoboken, New Jersey: Wiley)
[2] Unger H G 1977 *Planar optical waveguides and fibers* (Oxford: Clarendon Press.)
[3] Sklyarov O K 2001 *Fiber optic networks and communication systems* (Moscow: SOLON-Press) (Series Library Engineer)
[4] Agrawal G P 2012 *Fiber-optic communication systems* (Hoboken, New Jersey: Wiley)
[5] Salib M *et al* 2004 *Intel Technology J.* **8** 143
[6] Zheltikov A M 2004 *Phys. Usp.* **47** 1205
[7] Tanaka Y *et al* 2005 *Phys. Lett.* **86** 081108
[8] Berzhansky V N *et al* 2012 *J. Magn. Soc. Jpn.* **36** 42
[9] Inoue M *et al* 2006 *J. Mater. Chem.* **16** 678
[10] Inoue M *et al* 2006 *J. Phys.* D **39** R151
[11] Fedyanin A A *et al* 2003 *J. Magn. Magn. Mat.* **96** 258
[12] Fedyanin A A *et al* 2004 *J. Magn. Magn. Mat.* **282** 256
[13] Kahl S and Grishin A M 2004 *Appl. Phys. Lett.* **84** 1438
[14] Chakravarty A *et al* 2011 *Phys. Rev.* B **84** 094202
[15] Inoue M *et al* 2008 *Iieice Trans. Electron.* **e91–c** 1630
[16] Gaiyan B *et al* 2012 *Optical Materials* **35** 252
[17] Vasiliev M *et al* 2008 *IEEE Trans. Magn.* **44** 323
[18] Zvezdin A K and Belotelov V I 2004 *Eur. Phys. J.* B **37** 479
[19] Goto T *et al* 2009 *Phys. Rev.* B **79** 125103
[20] Vinogradov A P *et al* 2010 *Phys. Usp.* **53** 243
[21] Kalish A N *et al* 2014 *Laser Phys.* **24** 094006
[22] Prokhorov A M *et al* 1984 *Sov. Phys. Usp.* **27** 339
[23] Belotelov V I *et al* 2010 *J. Exp. Theor. Phys.* **137** 932
[24] Kreilkamp L E *et al* 2013 *Phys. Rev.* X **3** 041019
[25] Ferreiro-Vila E *et al* 2013 *Opt. Express* **21** 4917
[26] Yeh P and Yariv A 1977 *J. Opt. Soc. Amer.* **67** 423
[27] Yariv A and Yeh P 1984 *Optical Waves in Crystals: Propagation and Control of Laser Radiation* (New York: Wiley)
[28] Yeh P and Yariv A 1976 *Optics Communications* **19** 427
[29] Inoue M *et al* 2006 *J. Mater. Chem.* **16** 678
[30] Johnson S G *et al* 2001 *Opt. Express* **9** 748
[31] Xu Y *et al* 2003 *Opt. Express* **11** 1039
[32] Mizrahi A and Schachter L 2004 *Opt. Express* **12** 3156
[33] Li L 2003 *J. Opt.* A*: Pure Appl. Opt.* **5** 345
[34] Zvezdin A K and Kotov V A 1997 *Modern Magnetooptic and Magnetooptical Materials* (Philadelphia: IOP)
[35] Khokhlov N E *et al* 2015 *J. Phys.* D*: Appl. Phys.* **48** 095001

# Deconvolution of complex spectra into components by the bee swarm algorithm

**R R Yagfarov[1], M E Sibgatullin[2], D Z Galimullin[3], D I Kamalova[2] and M Kh Salakhov[2,4]**

[1]Innopolis, Innopolis, University St. 1, 420500
[2]Kazan, Kazan Federal University, Kremlyovskaya St. 18, 420008
[3]Kazan, Institute of Economics, Management and Law, Moscow St. 42, 420008
[4]Kazan, Academy of Sciences of the Republic of Tatarstan, Bauman St. 20, 420008

E-mail: goodraff@gmail.com

**Abstract.** The bee swarm algorithm is adapted for the solution of the problem of deconvolution of complex spectral contours into components. Comparison of biological concepts relating to the behaviour of bees in a colony and mathematical concepts relating to the quality of the obtained solutions is carried out (mean square error, random solutions in the each iteration). Model experiments, which have been realized on the example of a signal representing a sum of three Lorentz contours of various intensity and half-width, confirm the efficiency of the offered approach.

## 1. Introduction

Complex contours with unresolved structure are often registered when carrying out spectroscopic research. The examples of these spectra are IR absorption spectra of polymers [1]. To obtain complete and reliable information about the structure and properties of a substance, methods of deconvolution of spectral contours into elementary components are necessary. The spectra belonging to a particular chemical compound or spectral contours arising from quantum transitions and broadened by different mechanisms are usually called elementary components.

The aim of this work is implementation and adaptation of the method of deconvolution of molecular FTIR spectra into components using the artificial bee colony algorithm (ABC). This algorithm is a stochastic method for global optimization of a residual function based on the patterns of the behavior of bees [2]. Solutions with the least mean square error are taken as "foragers", whereas randomly generated solutions at the each iteration play a role of “scout bees”. "Flying out from the hive to harvest nectar, return to the hive and data processing" occur in one iteration. In order to assess the effectiveness of the bee algorithm for solving the problem of deconvolution, modeling experiments were carried out.

## 2. Swarm intelligence algorithm for decomposition of spectral contours

In this work the method of a swarm of bees was adapted to the solution of the problem of deconvolution of spectral contours into components. This algorithm simulates the behavior of bees in the natural environment. The idea of the bee algorithm is that in search of nectar all bees inspect at each step both high profitable nectar sources and the nectar sources in the vicinity of the high profitable ones. This allows, firstly, varying the population of solutions in subsequent iterations and, secondly, increasing the probability of finding solutions close to the optimum [3]. At first, several scout bees fly out from the hive in random directions to find nectar sources. After some time, the bees return to the hive, and in a special way inform the whole hive, where and how much nectar they have found. After that, another type of bees (foragers) fly to the found sites. The more nectar is expected to be found in the given site, the more bees fly in this direction, while scouts fly out again to look for other sites. After that the process is repeated. The position of global extremum corresponds to the site with the greatest amount of nectar, this section being the only one. This means that other sites contain a smaller amount of nectar.

The basic concepts of the algorithm are nectar source (flower, site), foragers (employed bees), scout bees. The source of nectar is characterized by its profitability determined by various parameters. Foragers are assigned to the sources of nectar. The number of bees on these sites is greater than on the other ones. The average number of scouts amounts to 5 - 10% of the swarm. After returning to the hive, the bees exchange information on the so-called closed dance area [4].

Let us consider in more detail the algorithm applied to the problem of deconvolution of spectral contours. Table 1 shows the comparison of biological and mathematical notions.

**Table 1.** Matching of classical and mathematical notions.

| **Classical term** | **Mathematical notion** |
|---|---|
| Nectar source (flower) | MSE (mean square error) $\sigma = \sqrt{\frac{1}{n}\sum_{i=1}^{n}(x_i - \bar{x})^2}$ |
| The most attractive sites | Solutions with the least MSE |
| Promising region | Solutions with a greater MSE than in attractive areas, but still quite promising |
| Bee | Solution as a sum of contours. In the case of a two-component Lorentz contour $P(\nu) = \frac{A_1}{\left(1 + \frac{(\nu - s_1)^2}{w_1^2}\right)} + \frac{A_2}{\left(1 + \frac{(\nu - s_2)^2}{w_2^2}\right)}$ |
| Foragers | A set of solutions which calculates MSE on the promising and attractive areas |
| Scouts | Random solutions at each iteration |
| Flying out, harvesting nectar, returning to the hive | One iteration |

The mean square error acts as a source of nectar. Search of the parameters of contour components is carried out in the space of dimension $N = n * m$, where $n$ is the number of contour components, $m$ is the number of unknown parameters of a component. Bees are the solutions obtained by adding up the estimated elementary components of a complex contour with certain parameters. Promising and most attractive sites are selected after ranking by the MSE value. Scouts are random solutions at each iteration of the algorithm. Foragers are the solutions calculated in the region of promising and attractive sites.

Figure 1 shows the block diagram of the algorithm. At the first stage a certain number of random initial sets of parameters of input contour components are generated. The number can be varied. Next, MSE is calculated, and ranking by the MSE value is performed. The $n$ solutions with the least objective function (MSE) are selected. So called promising sites $m$ are selected, where the objective function (MSE) is greater than in the best sites, but these sections can also be used for the search. After selecting the best and promising solutions at the given iteration, MSE in the domain of these solutions is calculated. The smaller is the MSE value, the more calculations are carried out in the domain of the given solution. Thus, the foragers are sent not exactly to the points where the best and the promising solutions have been found, but to the vicinity of these points. Then, ranking by MSE value is repeated, but now it is carried out not only among the scouts (random solutions), but also among all the bees (foragers and scouts). The algorithm is repeated until one of the stopping criteria is reached. Moreover, the parameters of the search area are changed at each iteration. The number of iterations or achievement of the deconvolution permissible error can be chosen as stopping criteria.

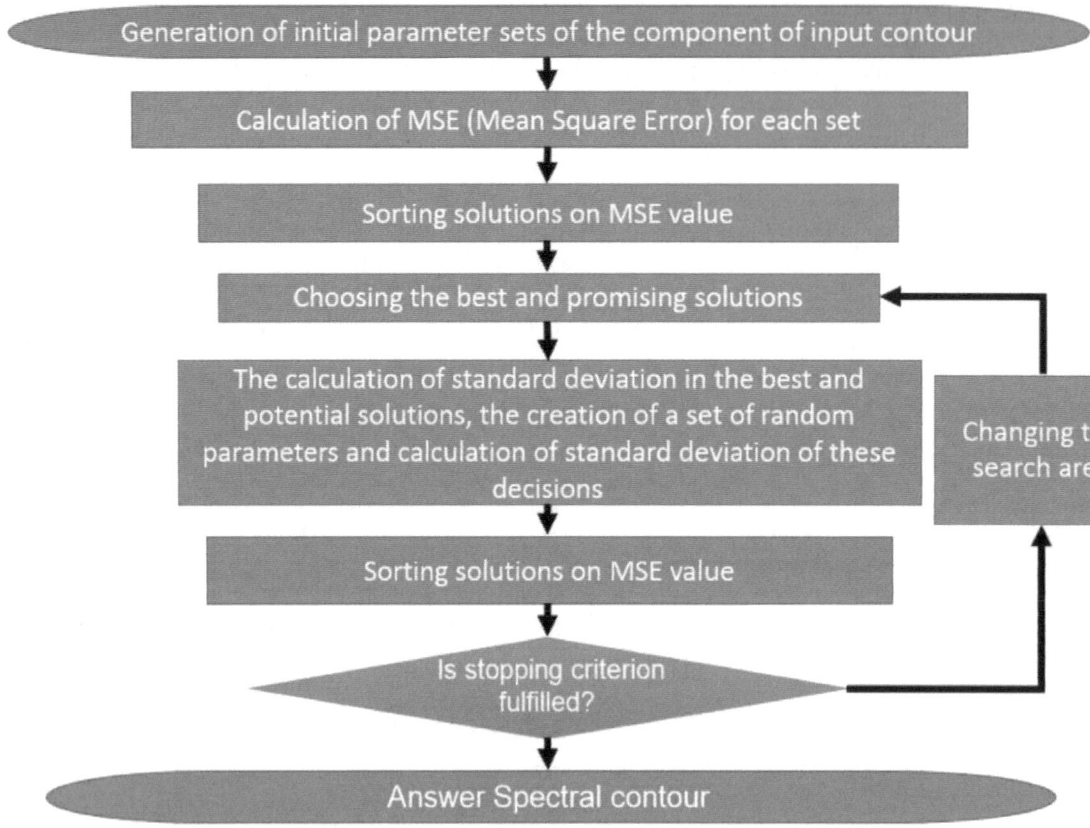


**Figure 1.** Block-diagram of the bee algorithm adapted to the problem.

In order to study the possibilities of applying the bee algorithm to the problem of deconvolution of complex contours, model experiments were carried out. The result of deconvolution of a three-component model using ABC is depicted in Figure 2. The results of deconvolution of the three-component model contour obtained by adding up three Lorentz contours are given in Table 2. These contours are characterized by the following parameters: $\boldsymbol{S_1} = \mathbf{2200}$, $\boldsymbol{A_1} = \mathbf{0.7}$, $\boldsymbol{W_1} = \mathbf{77}$ (first contour); $\boldsymbol{S_2} = \mathbf{2250}$, $\boldsymbol{A_2} = \mathbf{2}$, $\boldsymbol{W_2} = \mathbf{35}$ (second contour); $\boldsymbol{S_3} = \mathbf{2350}$, $\boldsymbol{A_3} = \mathbf{0.5}$, $\boldsymbol{W_3} = \mathbf{80}$ (third contour), where $\boldsymbol{S}$ is the position of the center, $\boldsymbol{A}$ is the intensity, $\boldsymbol{W}$ is the half-width. Deconvolution of a complex contour into components was performed using the ABC method and the least square method (OLS). The position of the component centers was taken as a priori information.

**Table 2.** Numerical values

| Method | $A_1$ | $A_2$ | $A_3$ | $W_1$ | $W_2$ | $W_3$ |
|---|---|---|---|---|---|---|
| OLS | -0,02 | 2,34 | 0,20 | 20681 | 57,93 | 215,39 |
| ABC | 0,69 | 1,99 | 0,50 | 76,99 | 35,07 | 80,22 |

As can see from figure 3, the use of OLS may lead to false results. The model contour is poorly resolved and represents the sum of three Lorentz contours with different intensities and half-widths. Each algorithm was run 2000 times. The number of false solutions for given model spectrum is amounted to 1% of all solutions in the case of ABC algorithm and to 35% in the case of OLS. Although the contour reconstituted by OLS is similar in the form to the original model contour, its components do not coincide with those of the original model contour. Therefore, in the case of multi-component spectral contours it is preferable to use the method of bee colonies adapted to solve the problem of deconvolution into components.

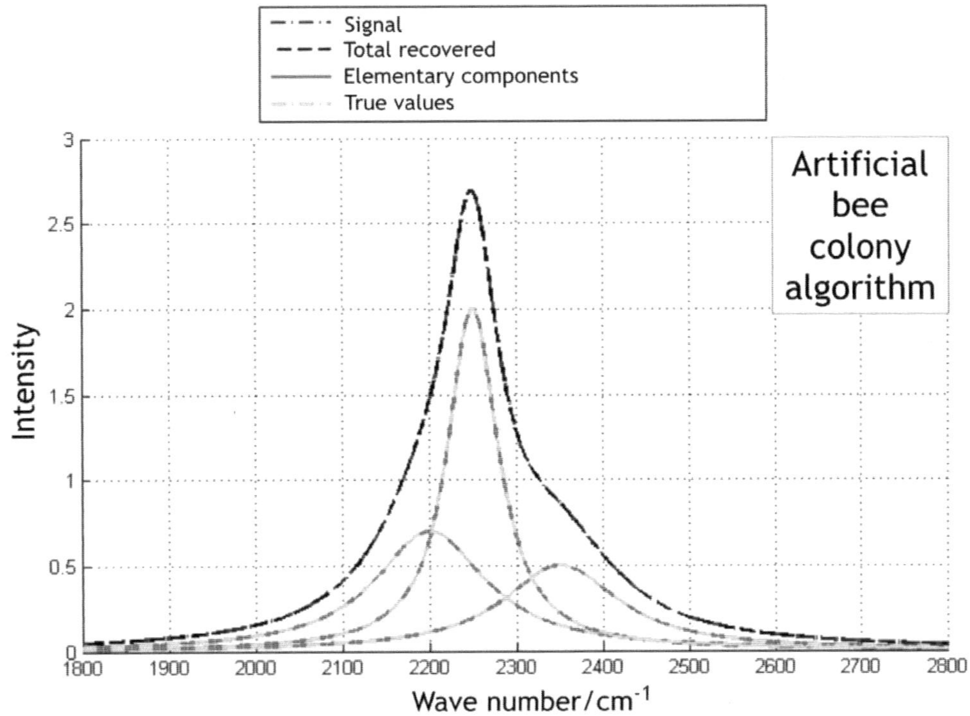


**Figure 2.** Deconvolution of the three-component model contour using ABC.

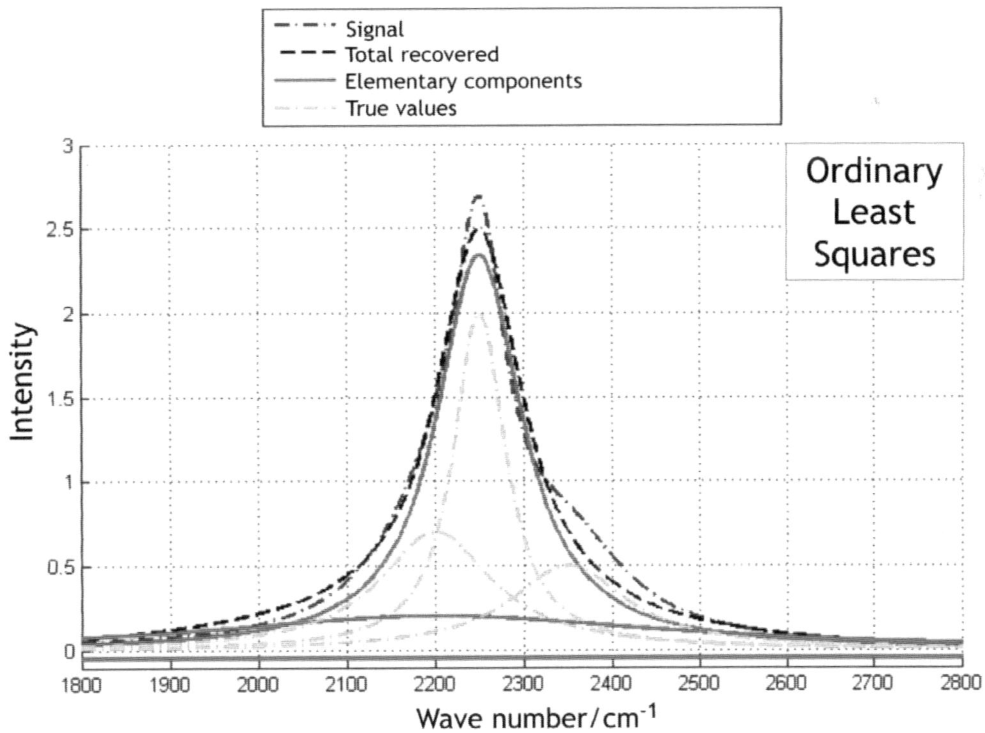


**Figure 3.** Deconvolution of the three-component model contour using OLS.

## 3. Conclusion

The artificial bee colony algorithm was implemented and adapted for processing model FT-IR absorption spectra. Model experiments on deconvolution of three-component spectral contours were performed. The possibility of using the artificial bee colony algorithm for solving the problem of deconvolution of poorly resolved spectral contours was shown. Also, comparison of the effectiveness of the proposed algorithm and least squares method was carried out. It is shown that in the case of a complex contour (in particular, of a three-component contour) using the least squares method increases the probability of obtaining false results when compared with the swarm intelligence algorithms.

**References**

[1] Kamalova D I, Remizov A B and Salakhov M Kh 2008 *Conformation probes in the study of local mobility of polymers* (Moscow: Fizmatkniga) (in Russian)

[2] Pham D T *et al* 2011 The Bees Algorithm–A Novel Tool for Complex Optimisation *Intelligent Production Machines and Systems - 2nd I* PROMS Virtual International Conference (3-14 July 2006)* ed D T Pham, E E Eldukhri and A J Soroka (Amsterdam: Elsevier) p 454

[3] Karaboga D 2005 An idea based on honey bee swarm for numerical optimization *Technical Report-tr06* (Erciyes University, Engineering faculty, Computer engineering department)

[4] Kureichik V M and Kazharov A A 2010 Graph coloring solution based on bee colony optimization *Izvestia SFedU. Engineering sciences* **113** (12) pp 30-36

# Coherent control for separation of the librational response through the ultrafast optically-heterodyne-detected optical-Kerr-effect (OHD-OKE)

**D K Zharkov, A G Shmelev, A V Leontyev, V G Nikiforov and V S Lobkov**

Kazan E. K. Zavoisky Physical-Technical Institute (KPhTI), Sibirsky tract, 10/7, 420029, Kazan, Russia

E-mail: dzharkov@list.ru

**Abstract**. We report the first experiment results on selective spectroscopy of librational response in ortho-dichlorobenzene at room temperature. By optimization of the controlling parameters we have achieved the suppression of the Raman-active vibrational and orientational responses. We have used the two-pulse excitation scheme. The controlling parameters were pulse duration, intensity and delay between two pump pulses.

## 1. Introduction

Femtosecond laser technique is one of the widest approaches for ultrafast processes studies. By using "pump-probe" method we can induce a certain excited state in the media by strong pump pulse and explore it by a weaker probe pulse. One of the methods is time-resolved ultrafast optical Kerr effect (OKE) spectroscopy of vibrational-rotational dynamics in liquids [1-3]. OKE signal consists of four different components [4,5]. The first, and the most rapid, is hyperpolarization response. The second one is rotation librational response. The third one is a longer orientational response. And the fourth is the vibrational response imposed on the whole spectrum (figure 1).

The frequency spectrum of the femtosecond laser pulse is very wide and allows to excite the above-mentioned responses. But a correct spectrum interpretation is a complex problem because of component superimposition. We have used the second pump pulse with certain parameters to suppress non-relevant responses and to distinguish the relevant one [6].

The delay times of vibrational-rotational responses are subpicosecond at room temperature. It can cause some difficulties for realization of multipulse experiment. Furthermore, laser coherent control of vibrational dynamics in chloroform and carbon tetrachloride is described in [7-12]. The cases of suppression and enhancement of the vibrational responses are realized for coherent laser control by the linearly polarized pulses. But the identical polarizations of the pump pulses don't allow to control the rotational dynamics.

The scenario of the suppression of the orientational and vibrational motions is realized in this work. Such a case allows us to perform the selective spectroscopy of the molecular librational motion response.

Journal of Physics: Conference Series **714** (2016) 012018 doi:10.1088/1742-6596/714/1/012018

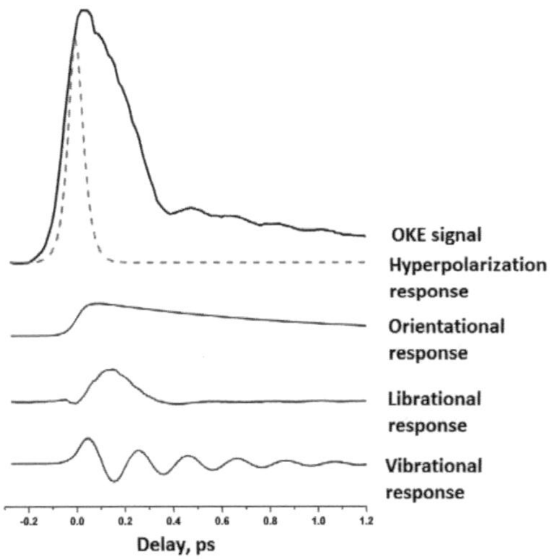


**Figure 1.** Components of the OKE signal.

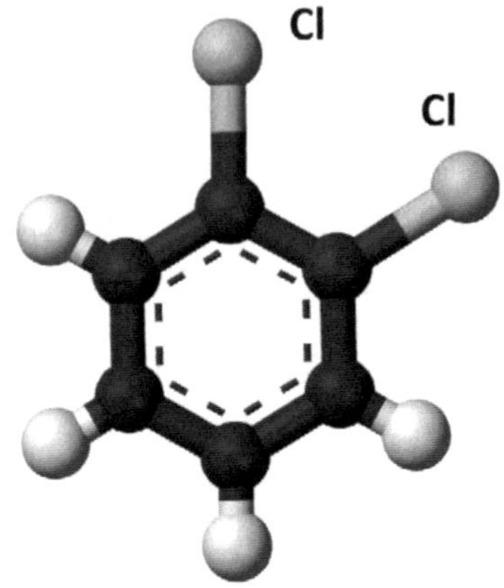


**Figure 2.** Molecule of ortho-dichlorobenzene.

## 2. Experiment

We have used a femtosecond Ti:sapphire laser with 800 nm wavelength. The pulse duration varied from 30 fs to 100 fs. The single pulse energy was $4\times10^{-9}$ J. Two pump pulses were polarized linearly, and mutually orthogonal. Pulse duration, intensity and delay were varied for optimal suppression and enhancement of the vibrational-rotational dynamics.

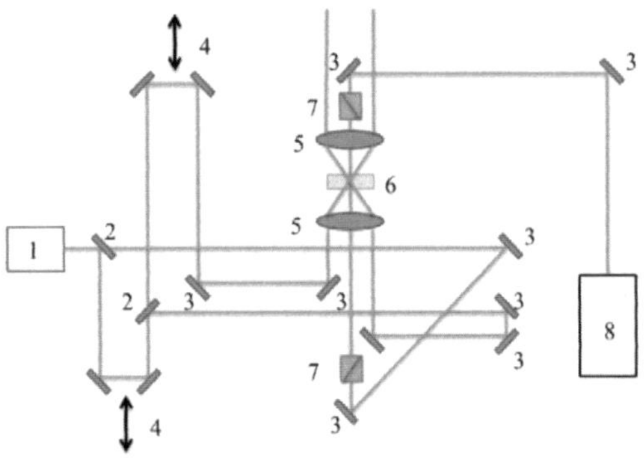


**Figure 3.** Principle scheme of the experimental setup. 1 – laser, 2 – beam splitter, 3 – mirror, 4 – optical delay line, 5 – lens, 6 – investigated liquid, 7 – polarizer, 8 – photomultiplier.

The laser radiation of the femtosecond laser was divided into three parts: two pump beams and the probe beam. The delay times of each pulse were controlled by the optical delay line. Two pump beams and the probe beam were focused on the investigated liquid by the lens. After passing the liquid cell the probe beam was separated from the pump beams by the aperture hole. For the reduction of the impact of the scattering pump on the probe beam the optical path was extended up to 5 meters. The state of the investigated liquid was characterized by the intensity change of the probe pulse passing through the first polarizer, the liquid and the second polarizer, which was crossed with the first one. So, the amplitude of the OKE signal is directly dependent on the induced anisotropy in the medium.

Ortho-dichlorobenzene (figure 2) was selected as the investigated liquid. It posesses all of the aforementioned responses. Ortho-dichlorobenzene molecule is asymmetric and flat. The dipole moment of this molecule is 2,14 D.

Journal of Physics: Conference Series **714** (2016) 012018 doi:10.1088/1742-6596/714/1/012018

## 3. Results

As we noted above, the OKE signal consists of the responses from electronic hyperpolarizability, Raman-active vibrations, orientation and libration motions. A typical spectrum of the single-pulse excitation OKE in liquid is depicted in figure 4. All four responses are represented. The main aim of this work was the implementation of selective spectroscopy of the librational response in ortho-dichlorobenzene. For the suppression of Raman-active vibrations we have used the second pump pulse with certain delay time. The polarization of the second pulse is the same as the polarization of the first pulse [8,13]. In [14] the possibility of the simultaneous suppression of the vibrational and orientational responses was theoretically illustrated. For this scenario the polarization of the second pulse must be orthogonal to that of the first.

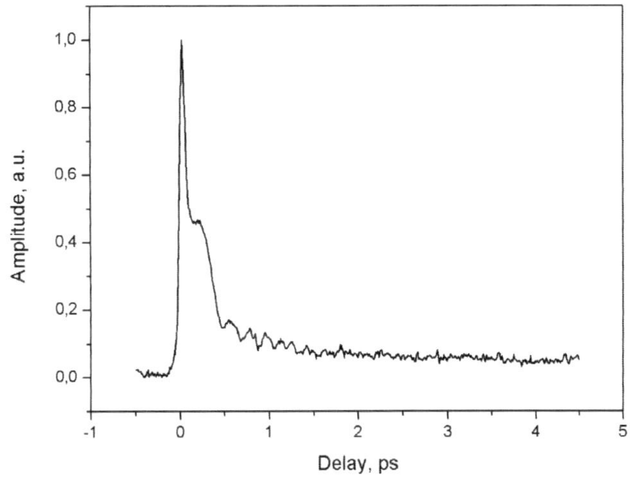


**Figure 4.** OKE signal in ortho-dichlorobenzene with single-pulse pumping.

The OKE signals from the first and the second pump pulses are depicted separately (The registered OKE signal is curve 3 in figures 5 and 6).

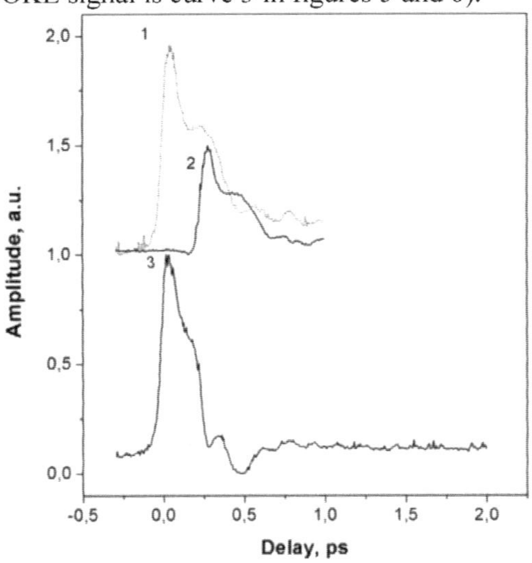


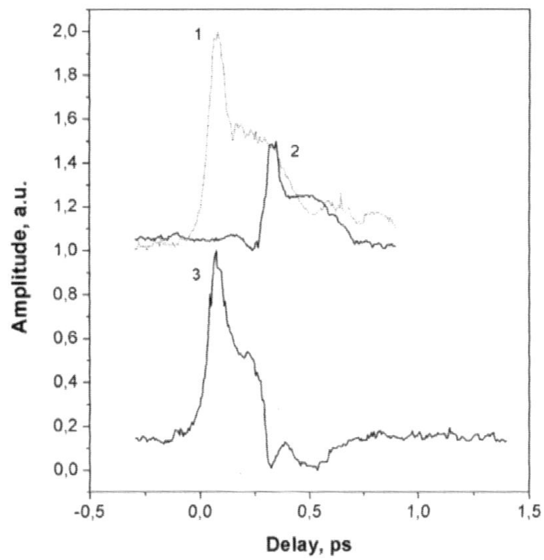


**Figure 5.** OKE signal in ortho-dichlorobenzene with pumping by the first pulse (1), the second pulse (2), the first and the second pulse simultaneously (3).

**Figure 6.** OKE signal in ortho-dichlorobenzene with pumping by the first pulse (1), the second pulse (2), the first and the second pulse simultaneously (3).

As it can be seen from the figure we were able to suppress orientational and Raman-active vibrational responses. In spectrum 3 we can clearly identify the librational response. We have successfully realized two cases of librational response selective spectroscopy with the following controlling parameters: pulses duration of 80 fs, the delay between pump pulses is 280 fs, intensity ratio is 1:2; pulses duration of 50 fs, the delay between pump pulses is 310 fs, intensity ratio is 1:2. The polarizations of the pump pulses were orthogonal to each other.

## 4. Conclusion

The first selective detection of the librational response is realized with OKE signal detection. The method is based on two-pulse laser coherent control of Raman-active intramolecular vibrations and rotational movements of molecules. The suppression of the non-relevant molecular responses is realized in the experiment. The final spectrum of OKE signal contains librational response only.

## References

[1] Ruhman S, Williams L R, Joly A G, Kohler B and Nelson K A 1987 *Journal of Physical Chemistry* **91** 2237–40
[2] McMorrow D, Lotshaw W T and Kenney-Wallace G 1988 *IEEE Journal of Quantum Electronics* **24** 443–54
[3] Righini R 1993 *Science – New York then Washington* **262** 1386
[4] Steffen T, Fourkas J T and Duppen K 1996 *The Journal of chemical physics* **105** 7364–82
[5] Nikiforov V G and Lobkov V S 2006 *Quantum Electronics* **36** 984
[6] Zharkov D K, Shmelev A G, Nikiforov V G and Lobkov V S 2014 *Journal of Physics: Conference Series* **572** 012043
[7] Nikiforov V G, Shmelev A G, Safiullin G M and Lobkov V S 2012 *Applied Physics Letters* **100** 081904
[8] Nikiforov V G, Shmelev A G, Safiullin G M and Lobkov V S 2012 *Quantum Electronics* **42** 332
[9] Nikiforov V G 2013 *Quantum Electronics* **43** 177
[10] Nikiforov V G, Lobkov V V and Samartsev V V 2015 *Laser Physics* **25** 075701
[11] Wiederrecht G P, Dougherty T P, Dhar L, Nelson K A, Leaird D E and Weiner A M 1995 *Physical Review* B **51** 916
[12] Weiner A M, Leaird D E, Wiederrecht G P and Nelson K A 1990 *Science* **247** 1317–9